Raphael Kozlovsky

Magnetotransport in 3D Topological Insulator Nanowires

Magnetotransport in 3D Topological Insulator Nanowires

Dissertation zur Erlangung des Doktorgrades der Naturwissenschaften (Dr. rer. nat.)
der Fakultät für Physik der Universität Regensburg
vorgelegt von

Raphael Kozlovsky

aus Freiburg im Breisgau
im Januar 2020

Die Arbeit wurde von Prof. Dr. Klaus Richter angeleitet.
Das Promotionsgesuch wurde am 26.11.2019 eingereicht.

Prüfungsausschuss:	Vorsitzender:	Prof. Dr. Dieter Weiss
	1. Gutachter:	Prof. Dr. Klaus Richter
	2. Gutachter:	Prof. Dr. Milena Grifoni
	weiterer Prüfer:	Prof. Dr. Christoph Lehner

Dissertationsreihe der Fakultät für Physik der Universität Regensburg, Band 54

Herausgegeben vom Präsidium des Alumnivereins der Physikalischen Fakultät:
Klaus Richter, Andreas Schäfer, Dieter Weiss

Raphael Kozlovsky

Magnetotransport in 3D Topological Insulator Nanowires

Universitätsverlag Regensburg

Bibliografische Informationen der Deutschen Bibliothek.
Die Deutsche Bibliothek verzeichnet diese Publikation
in der Deutschen Nationalbibliografie. Detailierte bibliografische Daten
sind im Internet über http://dnb.ddb.de abrufbar.

1. Auflage 2020

Leibnizstraße 13, 93055 Regensburg

Konzeption: Thomas Geiger
Umschlagentwurf: Franz Stadler, Designcooperative Nittenau eG
Layout: Raphael Kozlovsky
Druck: Docupoint, Magdeburg
ISBN: 978-3-86845-161-0

Weitere Informationen zum Verlagsprogramm erhalten Sie unter:
www.univerlag-regensburg.de

Abstract

Three-dimensional topological insulator (3DTI) nanowires host topologically protected surface states wrapped around an insulating bulk. In this dissertation, their transport characteristics in the presence of external electric and magnetic fields are investigated using effective surface Dirac Hamiltonians. In the first part, cylindrical nanowires are considered, which are made of the 3DTI strained mercury telluride (HgTe). A magnetic field along the wire axis leads to Aharonov-Bohm type conductance oscillations which signal surface transport, though alone cannot prove the underlying non-trivial topology of HgTe. Hence, we present an in-depth analysis of the gate-voltage dependent conductance oscillations which reveals the non-trivial topology, and gives insights about the effect of inhomogeneous charge carrier densities around the wire. In the second part of this dissertation, we go beyond the cylindrical geometry and consider shaped (tapered, curved) 3DTI nanowires. While their conductance in perpendicular magnetic fields is in general quantized due to higher-order topological hinge states, it depends on the precise wire geometry in coaxial magnetic fields, allowing to access several distinct transport regimes. For instance, for rotationally symmetric nanowires with varying radius such as a truncated cone, a coaxial magnetic field leads to a spatial variation of the enclosed magnetic flux, giving rise to a non-trivial mass potential along the wire direction. For the conical geometry, this mass potential leads to a conductance governed by transmission through Dirac Landau levels. Upon smoothing the cone, it is possible to enter the Coulomb blockade regime, where the cone acts as a quantum magnetic bottle confining surface Dirac electrons. Apart from truncated cones one can imagine countless other interesting wire geometries (with and without rotational symmetry) which grant access to a wealth of intriguing quantum transport phenomena, and may serve as building blocks for new types of Dirac electron optic setups.

Contents

1 Introduction

A major discovery in the research field of *topological states of matter*, nowadays one of the busiest subdisciplines in condensed matter physics, occurred in 2005 when the quantum spin Hall effect (QSHE) was discovered [1–3]. The roots of topology in condensed matter physics reach, however, far back. Topological ideas are very useful in explaining unusual physical phenomena, most prominently the quantum Hall effect (QHE) discovered experimentally by Klaus von Klitzing in 1980 [4]. For his discovery, a quantized Hall conductance in a two-dimensional (2D) electron gas at low temperatures and strong magnetic fields, he was rewarded with the Nobel prize in 1985, although the connection to topological ideas was not established yet.

Topology is a discipline of mathematics which is concerned with properties of objects which remain unchanged under continuous transformations, such as stretching or bending, but not tearing or gluing. One tangible example of such a property is the so-called genus of a closed surface, which corresponds to its number of holes. For instance, a sphere has genus zero whereas a torus has genus one. Properties as the genus, which are invariant under continuous transformations, are called *topological invariants*. Take, for instance, the torus and deform it into a tea cup by stretching and bending. The tea cup has still one hole (located at the handle), hence the genus is unchanged. The torus and the tea cup are said to be *topologically equivalent*, in the sense that their topological property, the genus, is equal. One can generalize the idea of the topology of a surface to the topology of manifolds. Then, by associating the Hamiltonian of a physical system with manifolds, it is possible to transfer the idea of topology from mathematics into the realm of physics.

A famous example of a topological invariant used in physics is the so-called *Chern number* C [5], which is closely related to the genus in the sense that their mathematical structure is very similar. The Gauss-Bonnet theorem states that the genus can be computed by solving the integral of the Gaussian curvature over a closed surface. By analogy, the Chern number is defined as the integral of the so-called Berry curvature over the Brillouin zone of a crystal (up to a prefactor). Accordingly, the Chern number C is a topological invariant like the genus and it takes discrete values. It turns out that the quantized Hall conductance that Klaus von Klitzing measured is given by Ce^2/h [6], where e is the elementary charge and h is the Planck constant. Astonishingly, the QHE allowed to measure the so-called Klitzing constant $R_K \equiv h/e^2$ with such an exceeding level of precision (one part per billion), that it was later used in metrology. The accuracy of the measurement can be readily explained exploiting arguments from topology: The Chern number C, which determines the Hall conductance, is a topological invariant and thus cannot change under continuous deformations of the system by definition. That is why the conductance is extremely robust against perturbations and does not depend on experimental details as, for instance, the amount of disorder.

It is this *robustness* which makes materials with topological properties so promising for future applications in electronics. The topological robustness can manifest itself in many different ways, depending on the physical system. As mentioned above, the resistance in the quantum Hall (QH) regime is not affected by the amount of disorder in the sample. The origin of this peculiar behavior is that scattering of electrons, which generates additional resistance and thus heat, is absent. The dynamics of the electrons is said to be *topological protected*. Modern days electronical devices, such as smartphones, consist of hundreds of billions of transistors which all produce heat. The heat development and consequently the large energy loss is one limiting factor when it comes to the further reduction of the size of transistors (which is crucial for increasing the computer power). Moreover, at the moment, the information and communication technology sector accounts for 5% to 9% of the total electricity consumption. Recent estimates suggest that this number could rise up to 20% in the next 10 years [7]. Hence, there is the urgent need for technology which operates much more efficiently in terms of energy consumption. The absence of scattering in certain topological materials, and hence conduction with vanishing resistance (apart from the contact resistance), may open a new pathway towards electronical devices with significantly reduced energy loss.

For the QHE, however, very low temperatures and strong magnetic fields are required, hence applications for electrical devices are not practicable. This brings us back to the QSHE, where no magnetic fields are needed and higher temperatures might be feasible without loosing the robustness of the conducting states. The QSHE appears in materials called topological insulators (TIs), which owe their name to the fact that they are insulating in the bulk but host conducting surface states which are topologically robust. Although, in principle, the topological robustness in TIs is similar to that of the QHE, their very nature is quite different. For instance, the surface states of a TI are only robust as long as time-reversal symmetry (TRS) is preserved. Also, one has to distinguish between 2D TIs, where the surface states are one-dimensional (1D) edge states, and three-dimensional (3D) TIs, first introduced in 2007 [8], where 2D surface states wrap around the entire sample. While the QSHE appears in the former, we will exclusively study the latter in this dissertation, which is motivated in the following.

The surface states of 3D topological insulators (3DTIs) have properties which are extremely interesting both for studying fundamental aspects of physics, as well as for future applications ranging from spintronics to quantum computation [9]. In the simplest case, their low-energy electronic structure is described by a 2D Dirac Hamiltonian, similar to that of graphene in the absence of inter-valley scattering. Each momentum along the Fermi contour has a well-defined spin state with a fixed angle between spin and momentum. Put in other words, the spin is locked to the momentum, usually referred to as *spin-momentum locking*. It follows that an electron which is backscattered (scattered by 180°), needs to flip its spin, which is only possible if the Hamiltonian comprises a perturbation which couples to the spin. However, such a perturbation, as for instance magnetic disorder, breaks TRS. Hence, we can conclude that direct backscattering is not allowed as long as TRS is preserved. This is an intuitive argument why localization by TRS-preserving disorder is absent in 3DTIs [10]. The robustness of the surface states and the important role of the spin explains the appeal of 3DTIs in quantum computation and spintronics. Moreover, the linear dispersion of the Dirac Hamiltonian allows to study fundamental aspects of relativistic quantum mechanics, such as the Klein paradox [11].

Many of the above mentioned properties of the surface states have been observed by angle-resolved photoemission spectroscopy (ARPES) (see for instance Refs. [12–14]). There is, however, the necessity to study 3DTIs in transport experiments in order to understand their macroscopic properties and to exploit their potential for future applications. Moreover, it is important to verify the ARPES results with an independent experimental method. However, extracting the properties of the surface states from transport experiments turns out to be quite difficult because of residual bulk conductivity. Although 3DTIs should, in principle, be gapped in the bulk, their bulk contribution to the conductance is often considerable due to residual carriers [15],[1] which masks the contribution to transport from the surface states. With ARPES, this problem does not occur because it is a purely surface sensitive method.

A simple but effective way of increasing the surface signal is to increase the surface to volume ratio, which can be done by building nanowires. On top of the reduced bulk contribution, nanowires open the pathway towards novel and interesting quantum phenomena: The Dirac-like surface states wrap coherently around the perimeter, giving rise to a non-trivial Berry phase originating from the spin-momentum locking mentioned above. Moreover, cross section sizes which

[1]For instance, due to charged Si vacancies in Bi_2Si_3 [16].

enable to thread a considerable magnetic flux through the core are possible [17],[2] allowing to study an interplay between Aharonov-Bohm physics, non-trivial Berry phase, and weak antilocalization [18]. All of the above mentioned phenomena can be studied in simple cylindrical or rectangular nanowires with constant cross section, which has been done experimentally, for instance, in Refs. [17, 19–24] and theoretically in Refs. [18, 25–27].

We extend this work by considering strained mercury telluride (HgTe) nanowires with a focus on asymmetric gating and Klein-tunneling in a joint theoretical and experimental study – and thereby discover a signature of the non-trivial topology [28]. Moreover, we leave the terrain of nanowires with constant cross section towards the much broader class of nanowires with arbitrary shape, and thereby open a new realm of novel fascinating quantum phenomena, especially if magnetic fields are involved. Depending on the shape of the nanowire and the magnetic field direction and strength, we will be able to study the QHE in curved space with Dirac electrons in this thesis, as well as extrinsic higher-order topological insulator phases [29–32]. Moreover, we will devise a *quantum magnetic bottle*, the quantum mechanical analog of a classical magnetic bottle, which allows to confine Dirac electrons leading to Coulomb blockade in transport. Such shaped nanowires literally offer a new dimension to play with compared to conventional planar electron optics, and could thus be envisaged as a building block for a new type of Dirac electron optic setups.

In the following, we give an overview of the topics discussed in this dissertation by providing a brief summary of each chapter.

In chapter 2, we supply the basic concepts necessary to understand subsequent chapters. We give a short introduction to the Landauer-Büttiker formalism and to Coulumb blockade. After a brief detour into the history of the research field topology in condensed matter physics, we discuss 3DTIs with a special focus on mercury telluride. We introduce the properties of the topologically non-trivial states residing on the surface of a 3DTI, and motivate the origin of their existence. Moreover, we introduce the numerical tools we apply to study the transport properties of 3DTI nanowires in later chapters. We thereby focus on tight-binding models which we obtain from effective continuum Hamiltonians, and which are the basis of our simulations.

Magnetotransport in nanowires with constant cross section are discussed in chapter 3. To this end, we set up the Hamiltonian for Dirac electrons on a 2D cylindrical surface. It turns out that a non-trivial Berry phase induced by the interplay between curvature and spin-momentum locking emerges in such system. We then discuss the tight-binding version of the Hamiltonian, and the appearance

[2] This is an advantage over carbon nanotubes, which typcially have a much smaller diameter.

of spurious solutions due to fermion doubling. In order to simulate realistic experiments, we explain how to implement efficiently correlated disorder, and how to realize leads resembling metallic contacts with many modes. Subsequently, we apply a coaxial magnetic field to the wire, which induces an Aharonov-Bohm phase. It turns out that for certain magnetic fluxes, the Aharonov-Bohm phase cancels the Berry phase and a special type of mode appears, which is protected from scattering – the so called perfectly transmitted mode. In the end of the chapter, we discuss the physics of a cylindrical 3DTI nanowire in perpendicular magnetic field, which is an example of a so-called extrinsic higher-order topological insulator.

In chapter 4, we present a joint experimental and theoretical work based on strained HgTe nanowires. The aim of this work is to provide a clear signature of the existence of Dirac-like states on the surface of these nanowires by means of transport experiments. To this end, we extend the model introduced in chapter 3 by adding a top gate which is used in the experiments to tune the Fermi energy. It turns out that the capacitance between top gate and nanowire plays a crucial role for obtaining the signature of the topologically non-trivial surface states. Hence, we present electrostatic simulations from which we obtain the capacitance, and reveal thereby that the top gate induces a strongly non-uniform electron density around the wire circumference. We show how the band structure is affected, and present a method to extract the spin-degeneracy of the surface states by a quantitative analysis of the conductance as a function of gate voltage.

Our first encounter with a shaped TI nanowire – namely a truncated TI nanocone – will occur in chapter 5. We start the chapter by introducing the effective continuum Hamiltonian describing Dirac electrons on such conical surfaces. We continue by presenting a powerful analytical tool to predict transport properties of nanocones in a coaxial magnetic field based on effective mass potentials. These mass potentials originate from the angular motion and incorporate the varying quantum confinement and the varying magnetic flux profile along the cone axis. The Dirac Hamiltonian is then transformed into a tight-binding version, and thereby a special numerical grid used for our transport simulations is presented. We finish the chapter by presenting magnetotransport simulations in weak and strong magnetic fields, revealing, among other things, resonant transport through Dirac Landau levels.

In chapter 6, we study arbitrarily shaped nanowires with nonzero Gaussian curvature. We start by focusing on rotationally symmetric nanowires in coaxial magnetic field. The rotational symmetry allows to derive a generalized version of the effective mass potentials of truncated nanocones. Exploiting these mass potentials, we present a *quantum magnetic bottle* – a device which allows to trap Dirac electrons with homogeneous magnetic fields, and to access intriguing

Coulomb blockade physics. Subsequently, we study an example of a nanowire which breaks rotational symmetry, and thereby develop a special non-uniform grid, which allows to devise tight-binding models for arbitrarily shaped nanowires.

Chapter 7 concludes this dissertation by providing a summary of our most important findings, and by giving an outlook toward possible future projects in this line of research.

2
Basic methodology

In this chapter we introduce the theoretical concepts and methods which will be important throughout this dissertation. We start with a brief introduction to the Landauer-Büttiker formalism and Coulomb blockade.

2.1 Mesoscopic physics and quantum transport

In this thesis, we deal with solid state systems in which the coherence length l_φ, which describes the length scale on which coherence is maintained and thus the dynamics is governed by quantum interference effects, is larger than the size of the system itself. The sizes we consider are, however, not on the atomic scale. We rather consider systems described by *mesoscopic* physics, which is situated in between the microscopic world of single atoms and the macroscopic world described by classical physics. The basic inquiry tool used in this thesis to learn more about the nature of those systems is based on quantum transport. We consider quantum transport experiments where the mesoscopic sample is placed between two or more electrodes which are assumed to be noninteracting and in thermal equilibrium. Then, a bias voltage V_b is applied between those electrodes and the current I through the sample acting as a scattering region is measured, which yields information about its properties. Thereby, we focus on the so-called linear response regime, where the applied bias voltage between the contacts is so small that the tunneling probability through the scatterer is not affected.

If the coupling to the leads is strong and the system can be treated as noninteracting (*i.e.* by a single particle Hamiltonian), the Landauer-Büttiker formalism [33–39] turns out to be an extremely successful tool in describing quantum transport, and we will heavily rely on it in the course of this thesis. It gives the relation between the conductance $G = I/V_b$ and the quantum mechanical transmission probability T through the scatterer. In the case of a two-terminal device, the formula reads

$$G = \frac{e^2}{h} \sum_{nm} T_{nm}(\epsilon_F), \tag{2.1}$$

where T_{nm} is the transmission probability between mode n in lead 1 and mode m in lead 2.[1] Note that even if the scattering region facilitates a perfect quantum mechanical transmission $T = 1$, the corresponding resistance does not vanish. The reason for this is the so-called contact resistance R_C, which we motivate briefly in the following. Suppose that there is only one mode in each lead, then Eq. (2.1) simplifies to $G = e^2T/h$, where $T \equiv T_{11}$, and the corresponding resistance is given by

$$R = \frac{1}{G} = \frac{h}{e^2}\frac{1}{T} = \frac{h}{e^2} + \frac{h}{e^2}\frac{1-T}{T}. \tag{2.2}$$

The first term on the right hand side of Eq. (2.2) is independent of the transmission probability T, *i.e.* independent of the scattering region, and can thus be associated with the contact resistance $R_C \equiv h/e^2$. The resistance of the scatterer is given by the second term $R_S \equiv h(1-T)/(e^2T)$.

Another transport regime, the so-called *Coulomb blockade* regime [40–42], can be accessed if the coupling to the electrodes is weak, *i.e.* if the tunneling rate Γ between the electrodes and the system under investigation fulfills $\hbar\Gamma \ll E_C$, where $E_C = e^2/(2C)$ is the so-called charging energy associated with the Coulomb energy that needs to be payed if an electron enters the central system.[2] Here, C is in general the capacitance between the system and the rest of the world, and e is the elementary charge. A description of Coulomb blockade is, in general, a complicated many-body problem that can be tackled in many different ways [42]. Throughout the course of this thesis, we will restrict ourselves to the so-called *constant interaction model*, which relies on the following two assumptions: First, all Coulomb interactions are incorporated by the *constant* charging energy E_C, and second, interactions do not modify the singe-particle spectrum of the central system. For details about the limitations of the constant interaction model we refer to Ref. [43]. Note that we do not include co-tunneling effects [44–46],[3] nor effects arising due to Kondo physics [47] (which become important for temperatures $k_BT \ll \hbar\Gamma$).

In the context of the Coulomb blockade regime, we will refer to the central system as quantum dot (QD), although it is not necessarily a zero-dimensional object. A so-called plunger gate, which couples electrostatically to the QD, can be used

[1] With spin-degenerate modes there is an additional factor of 2 in Eq. (2.1).

[2] The weak tunneling rate ensures that electrons are more or less localized in the center in between the two electrodes and thus feel the Coulomb repulsion of other electrons. Larger tunneling rates enable transport through delocalized states without having to pay charging energy.

[3] Co-tunneling describes tunneling processes through virtual charge states, where the electron resides on the dot for such a brief period of time that the energy uncertainty is large enough to pay for the energy deficit (the charging energy E_C). Co-tunneling is captured by second-order perturbation terms of the tunneling Hamiltonian.

to tune its electrostatic energy. For the rest of this thesis, we assume that the plunger gate is the only object with significant capacitive coupling to the QD. Hence, the capacitance C is solely determined by gate, dot, and the dielectric materials in between both.

In the following, we briefly introduce the constant-interaction model. Due to the week coupling Γ between QD and electrodes, the number of electrons N residing on the dot is well-defined, and its electrostatic energy is given by

$$E(N, V_g) = \frac{(Ne)^2}{2C} - V_g N e, \tag{2.3}$$

where V_g is the gate voltage of the plunger gate. The first term of Eq. (2.3) accounts for the Coulomb energy of N electrons on the dot, while the second term represents the electrostatic potential induced by the gate. For a fixed gate voltage, the number of electrons residing on the QD can be found by minimizing Eq. (2.3) for integer N. An interesting situation occurs when the gate voltage V_g is tuned such that $E(N, V_g) = E(N+1, V_g)$, which corresponds to a degenerate energy state of the QD, where the number of electrons can freely fluctuate between N and $N+1$. The gate voltage which fulfills this condition is given by $V_g = (N + 0.5)e/C$, suggesting that the energetically favorable situation corresponds to a charge of $N + 0.5$ electrons. For a gate voltage exactly in between those degenerate points, *i.e.* $\tilde{V}_g = Ne/C$, adding an extra electron costs an energy $E(N+1, \tilde{V}_g) - E(N, \tilde{V}_g) = e^2/(2C) \equiv E_C$. For current to flow, an electron has to tunnel through the first barrier, temporarily reside on the QD, and then tunnel out through the other barrier into the electrode. This requires that the charge can fluctuate between Ne and $(N+1)e$, from which we can deduce that for gate voltages $V_g \neq (N+0.5)e/C$ the conductance is suppressed since charge fluctuations require to pay Coulomb energy. This is only true if the required Coulomb energy is not provided by thermal excitations or bias voltage, *i.e.* as long as temperature and bias voltage fulfill $k_B T, eV_b \ll E_C$.

If the single particle level spacing of the QD is small compared to the charging energy, *i.e.* $\Delta\epsilon \ll E_C$, we expect – using arguments from above – periodically occurring maxima in the conductance G as a function of gate voltage V_g with a spacing $\Delta V_g = e/C$, which can be viewed as the "traditional" Coulomb blockade effect. In this thesis we favor, however, the situation where $\Delta\epsilon$ is on the same order as E_C. In this regime, the set of energies required for consecutive addition of electrons, which is often referred to as the *addition spectrum*, yields valuable information about the single particle spectrum.

The Coulomb blockade regime will be important in Sec. 6.2, while the capacitance between system and plunger gate will already be relevant in Ch. 4.

2.2 Topology in condensed matter physics

Nowadays, topological concepts are ubiquitous in condensed matter physics. In the following, we outline how topology has become such an important tool in this research field by giving a brief, and by no means exhaustive, historical overview of the most important discoveries.

We start with J. M. Kosterlitz and D. J. Thouless who obtained the Nobel prize in 2016, for, *inter alia*, their seminal work on topological phase transitions they pursued in the beginning of the '70s [48]. They discovered a new type of phase transition in 2D systems, nowadays known as the Kosterlitz-Thouless transition, which occurs between a quasi-ordered phase characterized by bound vortex pairs, and a disordered phase of free vortex and antivortex plasma. Since this transition is not accompanied by a spontaneous symmetry breaking, the Mermin-Wagner theorem [49] is not violated.[4] The Nobel prize for Kosterlitz and Thouless was shared with D. Haldane, who was honored for his work on quantum spin chains he conducted in the early '80s, which show non-trivial topological effects if the chain consists of integer spins [50–52].

Another milestone in the field of topology in condensed matter physics was the discovery of the so-called TKNN invariant by D. J. Thouless, M. Kohmoto, M. P. Nightingale, and M. den Nijs in 1982 [6], who explained the exceptionally accurate quantization of the Hall conductivity σ_{xy} in the integer QHE by showing that the quantization is topological in its nature, meaning that it is robust against any continuous deformation of the underlying band structure. Shortly after, B. Simons related the TKNN invariant to the more general class of Chern numbers [53], which can be used to classify distinct topological phases of certain systems with broken TRS. It turned out that those Chern numbers can be computed with the Berry curvature [54], which was introduced in its full generality by M. V. Berry in 1984. Assuming that a physical system depends on the parameters $\boldsymbol{R} = (R_1, R_2, ..., R_N)$ and is described by the Hamiltonian $H(\boldsymbol{R})$ with eigenstates $|\Psi(\boldsymbol{R})\rangle$, the Berry curvature is defined as

$$\boldsymbol{\Omega}(\boldsymbol{R}) = \mathrm{i}\nabla_{\boldsymbol{R}} \times \langle\Psi(\boldsymbol{R})|\,\nabla_{\boldsymbol{R}}\,|\Psi(\boldsymbol{R})\rangle\,, \tag{2.4}$$

where $\times$ is an n-dimensional generalization of the cross product. For Bloch systems [55] described by eigenstates in a Brillouin zone spanned by the crystal wavevector $\boldsymbol{k}$, the Chern number is computed by integrating over the Berry curvature $\boldsymbol{\Omega}(\boldsymbol{k})$ expressed in terms of the crystal wavevector, and summing over all occupied

[4]The Mermin-Wagner theorem states that continuous symmetries cannot be spontaneously broken in 2D (and 1D) systems at finite temperatures and finite range exchange interactions. Intuitively speaking, there are not enough interacting neighbors in 1D and 2D to stabilize a ground state which breaks the continous symmetry.

bands. Equation (2.4) can also be used to compute the so-called Berry phase, which appears if a quantum mechanical state evolves adiabatically along a closed parameter path $\mathcal{C}$, and returns to its original state only up to a phase factor, the Berry phase factor $\gamma_{\mathcal{C}}$. It is derived from the Berry curvature via

$$\gamma_{\mathcal{C}} = \int_{\mathcal{S}} \mathrm{d}\boldsymbol{S} \cdot \boldsymbol{\Omega}(\boldsymbol{R}), \tag{2.5}$$

where $\mathcal{S}$ is the surface enclosed by $\mathcal{C}$. We will see, for instance in Ch. 3, that such a Berry phase can have strong implications for the spectrum of a TI nanowire.

The Chern number is only one example of a topological invariant. In 1997, A. Altland and M. R. Zirnbauer introduced a classification scheme for generic Hamiltonians in the context of random matrix theory [56], which was transfered to the realm of topology in 2008 by A. P. Schnyder et al. [57]. This classification scheme if often called the "tenfold way"[5] and can be viewed as the "periodic table" of topological invariants. It classifies generic Hamiltonians according to their dimensionality and their symmetries, which comprise TRS, particle-hole symmetry, and sublattice symmetry. Here, the term generic means that all other symmetries are broken.[6] For instance, three dimensional systems where TRS squares to -1 (which is the case for spin-1/2 systems), and where particle-hole and sublattice symmetry are broken, are classified with the AII Cartan label. The corresponding topological invariant is described by $\mathbb{Z}_2$, a cyclic group consisting of only two elements, 0 (topologically trivial) and 1 (topologically non-trivial). This is the class of 3DTIs, the materials from which the nanowires we study in this thesis are made of.

A breakthrough in 2D materials occurred in 2004, when graphene was first successfully exfoliated by using a method based on scotch tape [61]. For this achievement, A. Geim and K. Novoselov were rewarded with the Nobel prize in 2010. In 2005, C. L. Kane and E. J. Mele proposed that at sufficiently low energies a single plane of graphene exhibits the QSHE [1, 62], where two counter-propagating spin-momentum locked modes travel at each edge of the sample. Those modes are degenerate and related by TRS (*i.e.* they are Kramers partners), and cross the spin-orbit coupling (SOC) induced gap in graphene. As mentioned in the introduction, they are topologically protected as long as TRS is preserved, which is directly related to the $\mathbb{Z}_2$ topological invariant. However, shortly after the SOC induced gap in graphene proved to be too small for the QSHE to be observable at realistic temperatures [63, 64]. Nonetheless, in 2007 M. König *et al.* observed the QSHE in 2D HgTe quantum wells (which have a much larger bulk gap than graphene) by measuring a quantized conductance $2e^2/h$ without any

[5]The tenfold way extends the threefold way developed by F. J. Dyson in 1962 [58].

[6]Recently, classification schemes including also spatial symmetries have been developed [59, 60].

magnetic field. Note that the existence of the QSHE in 2D HgTe quantum wells had been predicted before by B. A. Bernevig *et al.* theoretically [2].

Systems that potentially host the QSHE are called 2D TIs. The appearance of this topologically non-trivial quantum spin Hall phase can be traced back to an inverted band ordering, a concept that we will elaborate on in the next section. The concept of 2D TIs was soon generalized to 3DTIs [8, 65, 66], which host 2D conducting surface states instead of 1D edge states.[7] In the next section, we will give intuitive arguments for the existence of those surface states and provide precise details about their very nature.

2.3 3D topological insulators

3DTIs form a class of materials that are insulating in the bulk but host conducting states on the surface when surrounded by a trivial insulator or vacuum. In the following, we introduce the simplest Hamiltonian describing those surface states and discuss their properties. Subsequently, we explain their origin, and discuss the 3DTI strained Mercury Telluride (HgTe).

2.3.1 Nature of the surface states of a 3D topological insulator

In the simplest case, the electronic states on a flat surface of a 3DTI are described by a 2D Dirac cone [10] given by

$$\hat{H} = \hbar v_F (\hat{k}_x \sigma_x + \hat{k}_y \sigma_y), \tag{2.6}$$

with Pauli matrices σ_x, σ_y, wavenumber operators $\hat{k}_x = -\mathrm{i}\partial_x$, $\hat{k}_y = -\mathrm{i}\partial_y$, and the Fermi velocity v_F. In the following, we will show that the spin of the eigenstates of Eq. (2.6) always follows their momentum $\boldsymbol{p} = \hbar\boldsymbol{k}$, which is referred to as spin-momentum locking. Spin-momentum locked states are often called *helical* [67]. By using the plane wave ansatz $\Psi_{k_x k_y}(x, y) \propto \exp(\mathrm{i}k_x x)\exp(\mathrm{i}k_y y)\chi$, where χ is a two-component spinor, Eq. (2.6) can be diagonalized yielding the Dirac dispersion

$$\epsilon_\pm(k_x, k_y) = \pm\hbar v_F \sqrt{k_x^2 + k_y^2} \tag{2.7}$$

[7] In general, an n-dimensional TI hosts (n-1)-dimensional topologically non-trivial surface states.

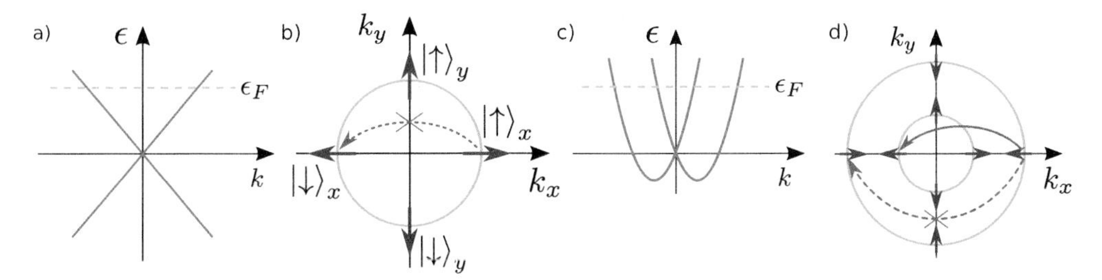

Figure 2.1: a) Dirac-like dispersion given by Eq. (2.7) as a function of k. b) Fermi surface at energy ϵ_F sketched in a) with a dashed line. Red arrows depict the spin direction of the eigenstates. Each arrow is labeled with the corresponding spinor (for instance, $|\uparrow\rangle_x$ labels an eigenvector of σ_x with positive eigenvalue). Scattering between Kramers partners is forbidden in the absence of TRS-breaking interactions, which is sketched by a dashed arrow with a cross. c) Dispersion of Schrödinger like metallic surface states with SOC, given by Eq. (2.12). d) The Fermi surface at an energy ϵ_F marked in c) consists of two circles, with opposite spin texture. Backscattering within each circle is still forbidden (gray dashed arrow with a cross). However, backscattering between the outer and the inner circle (and vice versa) is allowed (gray solid arrow).

with the corresponding spinor functions

$$\chi_+ = \frac{1}{\sqrt{2}} \begin{pmatrix} 1 \\ \frac{k_x + \mathrm{i}k_y}{k} \end{pmatrix} = \frac{1}{\sqrt{2}} \begin{pmatrix} 1 \\ \mathrm{e}^{\mathrm{i}\varphi} \end{pmatrix} \tag{2.8}$$

$$\chi_- = \frac{1}{\sqrt{2}} \begin{pmatrix} 1 \\ -\frac{k_x + \mathrm{i}k_y}{k} \end{pmatrix} = \frac{1}{\sqrt{2}} \begin{pmatrix} 1 \\ -\mathrm{e}^{\mathrm{i}\varphi} \end{pmatrix}, \tag{2.9}$$

where $\varphi \equiv \arctan(k_y/k_x)$. A standard parametrization of a two-component spinor in the Bloch sphere reads

$$\chi(\theta, \phi) = \cos\left(\frac{\theta}{2}\right) \begin{pmatrix} 1 \\ 0 \end{pmatrix} + \mathrm{e}^{\mathrm{i}\phi} \sin\left(\frac{\theta}{2}\right) \begin{pmatrix} 0 \\ 1 \end{pmatrix}, \tag{2.10}$$

where θ is the polar angle and ϕ the azimuthal angle. Comparing the eigenspinors χ_+ and χ_- with Eq. (2.10), it is evident that $\chi_+ = \chi(\theta = \pi/2, \phi = \varphi)$ and $\chi_- = \chi(\theta = \pi/2, \phi = \varphi + \pi)$. It directly follows that both spinors are confined in the xy-plane and that χ_+ is always parallel to the momentum while χ_- is antiparallel.

Figure 2.1 a) shows the linear Dirac dispersion Eq. (2.7) as a function of $k = \sqrt{k_x^2 + k_y^2}$, and Fig. 2.1 b) depicts the corresponding Fermi circle at an energy $\epsilon = \epsilon_F$. The spins of the eigenstates are sketched with red arrows. Due to spin-momentum locking, a spinor at wavenumber k is orthogonal to the spinor at wavenumber $-k$, *i.e.* counter propagating modes are orthogonal. We can conclude that direct backscattering is forbidden for disorder that does not couple to the spin. Actually, a general statement that can be proven for 3DTIs is that direct backscattering is forbidden as long as TRS is preserved [67]. Also, the scattering amplitude is weighted with the overlap between initial and final spinors, and thus scattering by large angles is in general suppressed. Note that although direct scattering by an angle of π is forbidden, it is allowed via multiple scattering events.

It is an important fact that the surface states of a 3DTI are fundamentally different from Schrödinger-like metallic surface states with SOC – no matter how strong the SOC. A possible form of a Hamiltonian $\hat{H}_S$ describing the latter is

$$\hat{H}_S = \frac{\hbar^2\left(\hat{k}_x^2 + \hat{k}_y^2\right)}{2m}\sigma_0 + \alpha\hbar(\hat{k}_x\sigma_x + \hat{k}_y\sigma_y), \tag{2.11}$$

which yields the dispersion

$$\epsilon_S^{\pm}(k_x, k_y) = \frac{\hbar^2\left(k_x^2 + k_y^2\right)}{2m} \pm \alpha\hbar\sqrt{k_x^2 + k_y^2}. \tag{2.12}$$

Here, σ_0 is the identity matrix in two dimensions, m is an effective mass, and α determines the strength of the SOC. The spinor structure of the eigenstates is exactly the same as for the Dirac Hamiltonian and is thus given by Eqs. (2.8) and (2.9). The corresponding dispersion as a function of k is shown in Fig. 2.1 c). No matter how large α or m, the Fermi surface at a given energy ϵ_F consists always of either zero or two closed curves [the latter is sketched in Fig. 2.1 d)] as opposed to a Fermi surface of the single Dirac cone, which consists of only one. Since the two branches have opposite spin-textures, there are always two counter propagating states that can scatter into each other, which is sketched by the solid gray arrow in Fig. 2.1 d). Hence, direct backscattering is not forbidden and localization is possible.

Generalized form of the surface Dirac Hamiltonian in flat space

A more general form of the 2D Dirac Hamiltonian on a flat surface, where the spin is confined to the surface but the angle θ between spin and momentum is a free parameter, is given by [25]

$$\begin{aligned}\hat{H} =& \hbar v_F[(\hat{n}_1\cdot\boldsymbol{\sigma})\cos\theta + (\hat{n}_2\cdot\boldsymbol{\sigma})\sin\theta](\hat{n}_1\cdot\boldsymbol{k}) \\ &+ \hbar v_F[(\hat{n}_2\cdot\boldsymbol{\sigma})\cos\theta - (\hat{n}_1\cdot\boldsymbol{\sigma})\sin\theta](\hat{n}_2\cdot\boldsymbol{k}),\end{aligned} \tag{2.13}$$

where $\hat{n}_1$ and $\hat{n}_2$ are two orthogonal vectors in the plane and $\boldsymbol{\sigma}$ is the vector of Pauli matrices. For $\hat{n}_1 = \hat{n}_x$, $\hat{n}_2 = \hat{n}_y$, and $\theta = 0$ we retrieve Eq. (2.6).[8]

[8] Another commonly used form is $\hat{H} = \hbar v_F(\hat{k}_x\sigma_y - \hat{k}_y\sigma_x)$ where $\theta = \pi/2$.

Let us emphasize that the properties of the surface states described above as well as all results in this thesis do not depend on which form of Eq. (2.13) is used since the relative orientation between spin and momentum is not relevant (only the locking between spin and momentum is important). The orientation becomes, however, relevant once additional spin-dependent terms are included, as for instance exchange or Zeeman coupling.

In general, the precise form of the Hamiltonian describing the states of a flat 3DTI surface in experiments can differ from Eq. (2.13). For instance, the velocity may depend on the direction [68–70], nonlinearities may appear away from the Dirac point [71, 72], or the spin polarization may be reduced [73, 74]. Also, the Fermi surface might consist of an odd number of curves larger than one [12]. However, the qualitative form of the main magnetotransport features we consider in this thesis are expected to be independent of these details [10].

Note that in reality, the surface of a 3DTI is always closed, *i.e.* it is never an infinite plane. In the upcoming chapter, we discuss what happens if the surface is a cylinder. In Chs. 5 and 6, we generalize the surface Dirac Hamiltonian and discuss curved 3DTI nanowires.

2.3.2 Origin of the topological surface states

In this section we give intuitive arguments for the existence of the topologically non-trivial states on surfaces/edges of TIs. We focus on 3DTIs, although the following arguments can also be applied to 2D TIs. It is instructive to first list a few TI materials found so far. The first experimental observation of a 3DTI phase was achieved in 2008 with $Bi_{0.9}Sb_{0.1}$ [12], followed in 2009 by Bi_2Se_3 [13, 75] and Bi_2Te_3 [14] (after the theoretical prediction by Zhang *et al.* [25]). In 2011, it was experimentally shown that strained HgTe is a 3DTI [76], which we will discuss in detail in the next section. All these crystals have in common that they consist of rather heavy elements (for example Hg has an atomic number of $Z = 80$). Consequently, the electrons in these crystals attain sufficient speeds such that relativistic effects play an important role. These effects are typically treated perturbatively to an order v^2/c^2 (where v is the electron velocity and c the speed of light) by adding the so-called mass-velocity term $\hat{H}_{\mathrm{mv}}$, the SOC term $\hat{H}_{\mathrm{SO}}$, and the Darwin term $\hat{H}_{\mathrm{D}}$ to the non-relativistic Hamiltonian. Depending on the material of the 3DTI, different correction terms prevail, but the crucial point is that all of them together lead to a so-called *band inversion* where the "normal" band ordering known from typical semiconductors or insulators is reversed. This effect is depicted in Fig. 2.2 b). The band structure of a trivial insulator is sketched on the left hand side, and its valence band is of p-type while the conduction band is of s-type around the Γ point of the Brillouin zone. On the right hand side, the

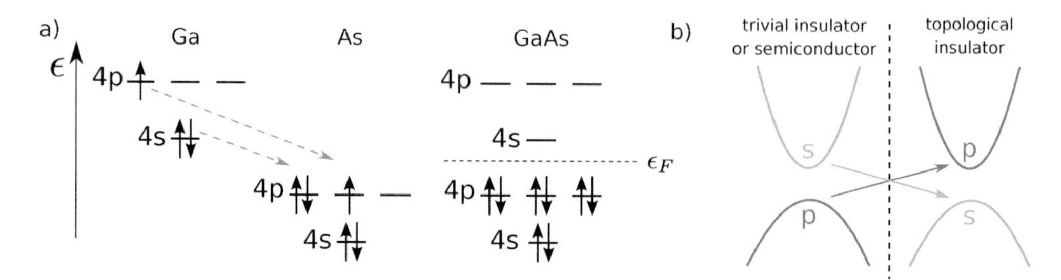

Figure 2.2: a) Sketch of the outer shell electron configuration of Ga and As. When forming GaAs, the 4s and 4p electrons of Ga pass over to As filling the p-shell which forms the valence band. The s-shell of Ga is left empty and forms the conduction band. b) Normal and inverted band ordering of a trivial and topological insulator. At the interface, the band gap closes and topological surface states form.

reversed band ordering of a topological insulator is shown. Bands are called of p- or s-type according to the atomic orbitals they are formed of. For clarification, let us consider as an example the electronic configuration of the semiconductor gallium arsenide (GaAs) in the atomic limit, sketched in Fig. 2.2 a). Gallium (Ga) has the electronic configuration $[Ar]3d^{10}4s^24p^1$ with an electronegativity of 1.6, while arsenic (As) has $[Ar]3d^{10}4s^24p^3$ with an electronegativity of 2.0. Bringing both together, As "swallows" the three $4s^24p^1$ electrons of Ga due to its larger electronegativity, which fills its p-shell. Hence, GaAs has a completely filled p-shell forming the valence band followed by an empty s-shell forming the conduction band. We will see in the next section how the p-shell rises energetically above the s-shell in HgTe leading to the inverted band structure of a 3DTI.

If a trivial insulator or semiconductor is brought into contact with a TI, the band structure has to continuously transform from one into the other. However, it is not possible to continuously transform an s-orbital into a p-orbital because they are orthogonal, and the s-orbital is symmetric while the p-orbital is antisymmetric under inversion symmetry. Hence, the only way out is that the s-type band of the trivial insulator goes down in energy while the p-type band goes up when approaching the TI, indicated by the arrows in Fig. 2.2 b). Consequently, the bands meet at the interface and the band gap is closed. These gap-crossing states are the much sought topological surface states (TSS).

A more formal way of explaining the existence of the topological surface states is based on topological invariants. In the context of TIs, a topological invariant is a discrete valued quantity which only depends on the topology of the wave functions in the Brillouin zone. It does not change if the parameters of the Hamiltonian are varied smoothly as long as the gap of the corresponding band structure is not closed. Consequently, the band gap must close at the interface of two materials with different values of the topological invariant. It is important to note that topological invariants can be computed only knowing the bulk wave functions, *i.e.* details about the surface structure are not important. Therefore, knowledge of the bulk Hamiltonian is enough to predict whether topological states exist on the surface. This is often referred to as *bulk-boundary correspondence.*

In this thesis, we work with materials which belong to the AII class in terms of the Altland-Zirnbauer table introduced in Sec. 2.2. The corresponding topological invariant $\mathbb{Z}_2$ is defined under the assumption that TRS is preserved, while particle-hole and sublattice symmetry are broken. We will, however, often break TRS by an external magnetic field in the course of this thesis. Let us emphasize that this does not imply a disappearance of the TSS. Clear signatures of Dirac surface states have been experimentally observed in 70 nm thick layers of strained HgTe (which belongs to the AII class) in perpendicular magnetic fields (up to 16 T were applied) [76]. This can be explained with the inverted band structure: as long as the magnetic field does not close the gap of the bulk band structure and reverses the band ordering back to trivial, TSS still exist. In a sense, the TSS are stabilized by the band gap. However, breaking TRS implies that the TSS are no longer necessarily protected from backscattering.

It turns out that in three dimensions four $\mathbb{Z}_2$ topological invariants are needed in the AII class to fully characterize the topological phase [8, 65, 66]. Three of them are called "weak" and one is called "strong". In this thesis we are only interested in the strong phase where the strong $\mathbb{Z}_2$ topological invariant is 1. Thus, when we talk about TIs we implicitly mean strong TIs. Computing $\mathbb{Z}_2$ topological invariants can be quite involved in general. However, depending on the system there are some simplifications, as for instance in the presence of inversion symmetry [77]. Finally, note that a TI surrounded by vacuum also hosts TSS, since vacuum has a trivial topology.

2.3.3 Strained HgTe

In this section, we will discuss the above mentioned band inversion by studying HgTe. The advantage of HgTe to Bi-based materials is its low bulk impurity concentration of $\approx 1 \times 10^{16}\,\mathrm{cm}^{-3}$, and its high mobility which can reach up to $4 \times 10^{5}\,\mathrm{cm}^2/\mathrm{Vs}$ when a $Cd_{0.7}Hg_{0.3}Te$ buffer layer is used [81].[9] Moreover, it is instructive to use HgTe as an example since it will be of central importance in Ch. 4 of this dissertation.

Although being often called a TI, bulk HgTe is actually a semimetal, *i.e.* it does not have a band gap. Conduction and valence bands are degenerate due to the cubic symmetry of the lattice and a quadratic band touching occurs at the

[9] Bi-based 3DTIs have a mobility on the order of $1 \times 10^{3}\,\mathrm{cm}^2/\mathrm{Vs}$.

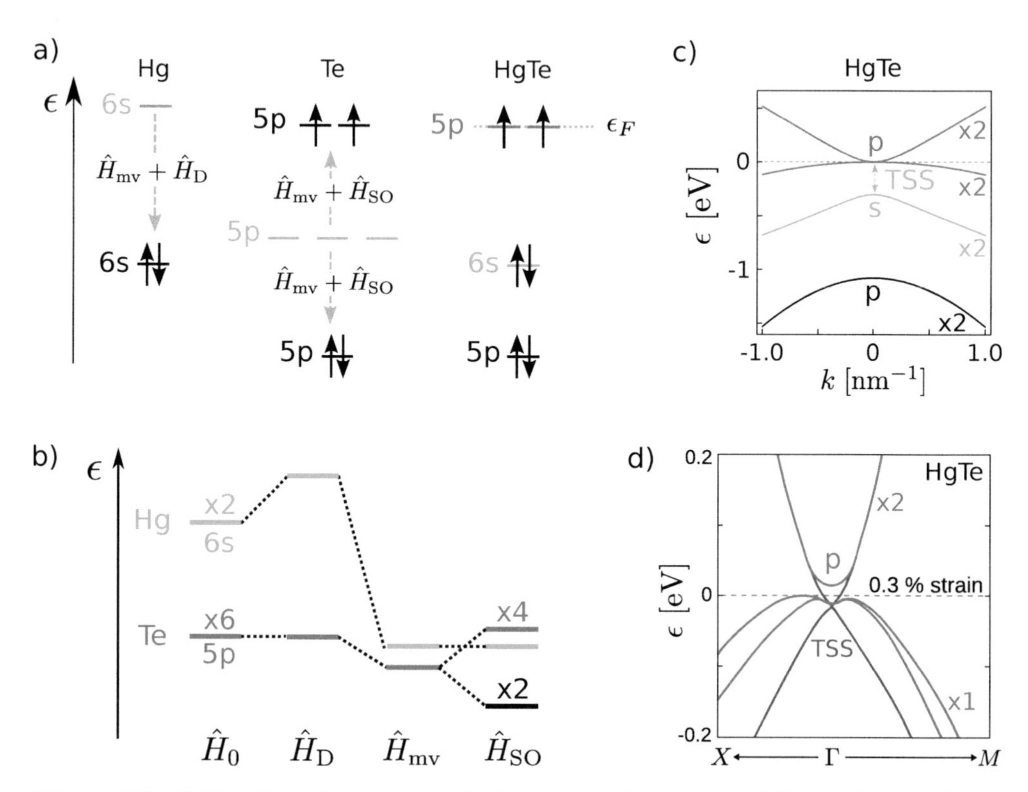

Figure 2.3: a) Sketch of the outer shell electron configuration of Hg and Te, and the shift in energy of the orbitals due to relativistic effects. The shift is sketched with gray arrows. Due to the relativistic corrections, part of the p-shell is energetically above the s-shell in HgTe, which reflects its inverted character. b) Individual contributions of the relativistic corrections to the Hg 6s-shell and Te 5p-shell. The six-fold degeneracy of the 5p-shell of Te is split by the SOC term into one four-fold (blue) and one two-fold degeneracy (black) [see also a)]. Adapted from Ref. [78]. c) Band structure of bulk HgTe. The four-fold degenerate 5p-shell of Te (after relativistic corrections) is occupied by two electrons [see a)]. The occupied orbitals form the valence and the empty orbitals the conduction bands (both depicted by blue curves). The 6s-shell of Hg forms the two-fold degenerate band below the valence band (green curve). Adapted from Ref. [2]. d) Tensile strain opens an indirect gap between conduction and valence band and splits the degeneracy of the valence band away from the Γ-point. The Dirac point of the TSS (red color) is located between the p- and s-type bands and is thus submerged in the valence bands. Nevertheless, there is an energy window (the band gap) where only TSS exist. This is where the non-trivial topology of HgTe is accessible in magnetotransport experiments. Sketch adapted from [79, 80].

$\Gamma-$point of the Brillouin zone, as can be seen in Fig. 2.3 c) at $\epsilon = 0$. Nevertheless, it has an inverted band structure and is topologically non-trivial as we will see shortly. Without a band gap the properties of the TSS are obscured by bulk states and can hardly be measured in experiments. However, a band gap can be opened at the Fermi energy [dashed line in Fig. 2.3 c)] in HgTe by forming a 2D quantum well, where a gap opens due to quantum confinement. This method led to the first experimental realization of a 2D TI achieved with $HgTe/Hg_{0.3}Cd_{0.7}Te$ quantum well structures [2, 3]. Alternatively, a band gap can be opened by applying uniaxial tensile strain to HgTe, which was predicted in 2007 by Fu *et al.* [65] and experimentally observed in 2011 by Brüne *et al.* [76]. Thereby, HgTe is grown on cadmium telluride (CdTe) which has a lattice constant that is 0.3% larger. The HgTe layer adopts the lateral lattice constant of CdTe, which breaks the cubic symmetry and opens a band gap. The strain induced gap opening is studied in detail in Refs. [79, 82, 83] and is sketched in Fig. 2.3 d).

Let us now elaborate on the topologically non-trivial character of HgTe. Figure 2.3 a) shows the electronic configuration of the outer electronic shells of Hg and Te. Due to the net contribution of the mass velocity term $\hat{H}_{\mathrm{mv}}$ and the Darwin term $\hat{H}_{\mathrm{D}}$, the 6s-orbital of Hg is strongly lowered in energy. The 5p-shell of Te instead is split by the SOC term and two of the 5p-orbitals (each orbital is two-fold spin-degenerate) rise above the 6s orbital of Hg. The individual contributions of the relativistic corrections are sketched in Fig. 2.3 b). In HgTe, the two electrons in the 5p-orbitals of Te whose energy was increased by the relativistic corrections form the p-type valence bands in Fig. 2.3 c) (blue color), while the two empty slots form the two-fold degenerate conduction bands. It is important to note that both the conduction and the valence band are of p-type. The two electrons in the 6s-orbital of Hg form the two-fold degenerate s-type band below the valence band (green color). Hence, the band inversion in HgTe is different from the one shown in Fig. 2.2 b) since it does not occur between conduction and valence band, but rather between conduction band and the band *below* the valence band. Using arguments about the appearance of TSS due to band inversion from the last section, we know that the TSS must be energetically located somewhere in between the p-type bands and s-type bands. As a consequence, the Dirac point is submerged in the valence bands.

By opening a gap with tensile strain, it can be shown that the Dirac point moves slightly towards the gap, but it is still submerged in the valence bands for a strain of 0.3% [79, 82], which is the experimentally feasible value. A sketch of the TSS partially submerged in the valence bands is shown in Fig. 2.3 d). The fact that the Dirac point of the TSS is not accessible in the band gap is actually a strong drawback of HgTe. In Ch. 3, we will see that the lowest mode in 3DTI nanowires, which is close to the Dirac point, is protected from scattering. Measuring this protection would constitute a direct prove of its topological nature. However, for experimentally feasible wire sizes, the lowest mode is submerged in the valence band and its transport contribution is thus masked and cannot be observed. Hence, one needs to develop different methods to prove the non-trivial topology of HgTe by means of transport experiments, which is presented in Ch. 4.

2.4 Quantum transport simulations with tight-binding models

Most of the numerical simulations presented in this thesis were performed with the Python software package *kwant* [84], which we will introduce in the following. *Kwant* is a software package with the purpose of simulating physical systems that can be described by single-particle tight-binding Hamiltonians. Both finite systems and infinitely large systems can be treated. Its main purpose, however, is to treat systems which consist of a finite scattering region to which semi-infinite periodic leads are attached. Those leads act as wave guides hosting plane waves entering and exiting the scattering region (as in the Landauer-Büttiker formalism, see Sec. 2.1). The leads correspond to contacts in a quantum transport experiment.

The default solver of *kwant* is based on the wave function matching method, which is described in detail in Ref. [85]. *Kwant* allows to compute many properties of tight-binding systems, as for instance, wave functions, the dispersion relation, the density of states, the scattering matrix, the current density, and the conductance (to name just a few).

2.4.1 Effective continuum Hamiltonians and tight-binding models

Kwant takes as an input a finite dimensional matrix representing a tight-binding model, which can, for instance, originate from an atomic description where each site corresponds to an atom. In our case, however, the tight-binding models originate from a finite-difference discretizations of effective low energy continuum Hamiltonians. In other words, we start from continuum Hamiltonians and solve the Dirac (or Schrödinger equation) on a numerical lattice. In the following, we briefly outline the procedure to derive such discretized Hamiltonians in matrix form from a continuum model, and give the connection to "atomic" tight-binding models. Thereby, we introduce the notation for tight-binding Hamiltonians that is used throughout this thesis.

As an example, consider a 1D toy model system of length L with periodic boundary conditions described by a continuum Hamiltonian linear in the momentum, *i.e.* $\hat{H} = v_F \sigma_x \hat{p}$. The Pauli matrix σ_x representing a spin or pseudo-spin degree of freedom ensures TRS.[10] The Hamiltonian in real space is given by $\hat{H} = v_F \sigma_x(-\mathrm{i}\hbar\partial_x)$ and we approximate the derivative with a finite number of difference quotients defined on an artificial lattice, often referred to as the *numerical grid.* The two-component spinor wave function $\Psi(x)$ is thereby defined on a finite number of lattice points N, and we use the shorthand notation $\Psi_j \equiv \Psi(x = ja)$, where a is the lattice spacing. The lattice index j takes integer values from $j = 1$ to $j = N$ and we use the definition that site 0 is equivalent to site N and site $N+1$ is equivalent to site 1 (due to the periodic boundary conditions). With those definitions, the momentum operator acting on the spinor yields in the discretized form

$$\hat{p}\Psi(x)\Big|_{x=x_j} = -\frac{\mathrm{i}\hbar}{2a}\left(\Psi_{j+1} - \Psi_{j-1}\right). \tag{2.14}$$

We use the symmetric finite difference approach above in order to obtain a Hermitian Hamiltonian matrix in the end. Using Eq. (2.14), the eigenvalue problem $\hat{H}\Psi = \epsilon\Psi$ defines a set of equations

$$v_F \sigma_x \frac{-\mathrm{i}\hbar}{2a}\left(\Psi_{j+1} - \Psi_{j-1}\right) = \epsilon\Psi_j, \tag{2.15}$$

which can be written in matrix form as

$$\underbrace{\begin{pmatrix} 0 & t & 0 & 0 & & \dots & t \\ -t & 0 & t & 0 & & \dots & 0 \\ 0 & -t & 0 & t & & \dots & 0 \\ \vdots & & & & & & \vdots \\ 0 & \dots & & & -t & 0 & t \\ -t & \dots & & & 0 & -t & 0 \end{pmatrix}}_{H} \begin{pmatrix} \Psi_1 \\ \Psi_2 \\ \vdots \\ \vdots \\ \Psi_{N-1} \\ \Psi_N \end{pmatrix} = \epsilon \begin{pmatrix} \Psi_1 \\ \Psi_2 \\ \vdots \\ \vdots \\ \Psi_{N-1} \\ \Psi_N \end{pmatrix}, \tag{2.16}$$

where $t = v_F \frac{\mathrm{i}\hbar}{2a}\sigma_x$. In the following, we use $\hat{H}$ to denote operators and H to denote matrices. It is the Hamiltonian matrix H which is required as an input by *kwant* (although there are routines within *kwant* which help to construct H starting from $\hat{H}$). Although the numerical lattice has nothing to do with an atomic lattice, H clearly has the form of a tight-binding Hamiltonian with complex nearest-neighbor hopping t and a spinor-like degree of freedom on each atom. The same procedure can be applied to the Schrödinger Hamiltonian $\hat{H}_{\mathrm{Schr}} = \hat{p}^2/(2m)$, see App. A.1 for details. There it is shown that the discretized version of $\hat{H}_{\mathrm{Schr}}$ corresponds exactly to the tight-binding Hamiltonian of an atomic chain with nearest-neighbor hopping.

[10] Note that this is not the Hamiltonian for a Dirac electron on a ring since the curvature-induced spin rotation (originating from spin-momentum locking) is not included. This spin rotation results in a π-Berry phase. We refer to Sec. 3.1 for details.

Writing the Hamiltonian in the form of Eq. (2.16) can become very cumbersome if there are more degrees of freedom. Thus, it is convenient to use a projection operator-based language (usually used for tight-binding models), *i.e.* to write H as

$$H = t \sum_j |j\rangle \langle j+1| + \text{h.c.} \tag{2.17}$$

The term $t\,|j\rangle\langle j+1|$ represents the hopping between site j and $j+1$ where t is the hopping amplitude (hopping integral), which is a complex matrix in our case, and h.c. is the short-hand notation for Hermitian conjugate. Mathematically, the term $t\,|j\rangle\langle j+1|$ simply sets an entry in row j and column $j+1$ in the matrix H. A more general version of a tight-binding Hamiltonian H_{TB} with an arbitrary number of dimensions and orbital degrees of freedom can be written in the form

$$H_{\mathrm{TB}} = \sum_{\boldsymbol{r},\boldsymbol{r}'} \sum_{\alpha,\alpha'} H_{\boldsymbol{r}\alpha\boldsymbol{r}'\alpha'} |\boldsymbol{r}\alpha\rangle \langle \boldsymbol{r}'\alpha'| , \tag{2.18}$$

where $\boldsymbol{r}$, $\boldsymbol{r}'$ label spatial degrees of freedom (the sites of the lattice) and α, α' label internal degrees of freedom (*e.g.* spin). In the upcoming chapters, we will make use of this notation.

Now, let us continue with the analysis of the 1D toy model. As we will see, it is instructive to solve the eigenvalue problem in its continuous form $\hat{H}\Psi = \epsilon_{\mathrm{con}}\Psi$ and to compare its eigenvalues to those of the discretized version given by Eq. (2.15). Using a plane wave ansatz for the spatial part of the wave function $\Psi(x) = \frac{1}{\sqrt{L}}\mathrm{e}^{ikx}\chi$ (translational invariance), with wavenumber k and spinor χ, the solution of the continuous problem is obtained by diagonalizing a 2x2 matrix, which results in the dispersion

$$\epsilon_{\mathrm{con}}(k) = \pm\hbar v_F |k|. \tag{2.19}$$

The allowed wavenumbers k are determined by the boundary condition $\mathrm{e}^{\mathrm{i}k(x+L)} = \mathrm{e}^{\mathrm{i}kx}$, which yields $k_l = \frac{2\pi}{L}l$, with $l \in \mathbb{Z}$. Returning to the discretized description of the problem [Eq. (2.15)], we use the same ansatz as before but on the lattice, *i.e.* $\Psi_j = \frac{1}{\sqrt{L}}\mathrm{e}^{ikja}\chi$. This leads to the equation

$$v_F \sigma_x \frac{-\mathrm{i}\hbar}{2a} \left[\mathrm{e}^{\mathrm{i}k(j+1)a} - \mathrm{e}^{\mathrm{i}k(j-1)a}\right] \chi = \epsilon_{\mathrm{dis}}\, \mathrm{e}^{\mathrm{i}kj}\chi, \tag{2.20}$$

which is equivalent to

$$v_F \sigma_x \hbar \frac{1}{a} \sin(ka)\chi = \epsilon_{\mathrm{dis}}\chi. \tag{2.21}$$

Equation (2.21) is again a 2x2-matrix equation which can be diagonalized and yields the dispersion for the discretized system

$$\epsilon_{\mathrm{dis}}(k) = \pm\hbar v_F \frac{1}{a} |\sin(ka)|. \tag{2.22}$$

As before, the allowed k-values are given by $k_l = \frac{2\pi}{L} l$. However, this time there is an additional constraint on the allowed l-values. Due to the structural equivalence of H to an atomic tight-binding model, it is clear that there is an "artificial" Brillouin zone associated with the numerical lattice. Thus, k_l can only take values in between $-\pi/a$ and π/a; adding other k_l-values gives redundant solutions. Note that for $ka \ll 1$, we have $\epsilon_{\text{con}} \approx \epsilon_{\text{dis}}$, *i.e.* in the large wave-length limit the electron does not feel the lattice (as expected). For larger ka, however, the electron starts to feel the lattice leading to a deviation between discretized and continuum solutions. In the next section, we will see that this deviation has strong repercussions for numerical simulations, also (surprisingly) in the low energy limit. This has to do with the fact that ϵ_{dis} does not increase monotonically, which leads us to the so-called *Fermion doubling*.

2.4.2 Fermion doubling

Fermion doubling is the appearance of spurious solutions whenever Dirac fermions are represented on a spatial lattice if the corresponding Hamiltonian is Hermitian. It has been known for decades in high-energy physics [86, 87]. A general formulation of the problem is given by the so-called Nielsen-Ninomiya theorem published in 1981 [88–90]. In condensed matter physics, it was not until 2004 that an ubiquitous interest in the Dirac equation emerged (and with it the problem of fermion doubling) with the experimental discovery of single layer graphene. Nowadays, there are countless condensed matter systems described by the Dirac equation, which can be attributed to the discovery of topological insulators by a large part. Accordingly, there is a vast demand in simulating Dirac materials in condensed matter physics. Note that unphysical fermion doubling does not appear in atomic tight-binding models (where each site corresponds to an atom), but it appears when discretizing effective continuum Hamiltonians. However, since computational costs for simulations with atomic tight-binding models are often too large to be feasible (typically the physical systems considered consist of a huge number of atoms), one often resorts to discretized versions of effective Dirac Hamiltonians despite the drawback of the spurious solutions. In the following, we will see how fermion doubling appears due to the central, three-point discretization of the first order spatial derivative of the 1D toy model introduced in the last section.

For simplicity, we consider the case $L \to \infty$. This yields a continuous spectrum for the discretized representation, which can be conveniently compared with the continuum model. A plot of the corresponding Eqs. (2.19) and (2.22) is shown in Fig. 2.4 b) in the first Brillouin zone. As already mentioned, for small $|k|$-values, the dispersion of the continuous system is well discribed by the discretised one. However, for $k = \pi/2a$, ϵ_{dis} reaches an extremum and then returns to zero

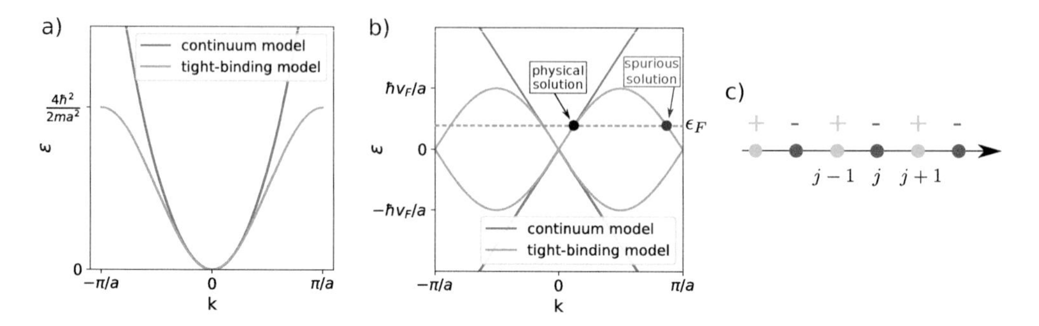

Figure 2.4: Continuum and the corresponding tight-binding band structures for a) the Schrödinger Hamiltonian [see App. A.1 and Eq. (A.4)] and b) the Dirac Hamiltonian. In the case of the Dirac Hamiltonian, there is a spurious solution for all Fermi energies within the band width. c) Spurious solution appearing in the discretized version of the Dirac Hamiltonian for $k = \pi/a$. With a symmetric finite difference scheme, this highly fluctuating state has zero energy.

energy at $k = \pi/a$. The consequence is that for every Fermi energy within the band (*i.e.* $-\hbar v_F/a < \epsilon_F < \hbar v_F/a$), there are twice as many fermions as for the continuous system, hence the term fermion doubling. This means that even at low energies where the finite difference approximation is supposed to work well there are always spurious solutions of the Dirac equation, which might have repercussions on the numerical simulation. We want to emphasize that this is in contrast to Schrödigner systems [see Fig. 2.4 a) and App. A.1], where no doubling occurs and thus a restriction to low energies is enough to capture the correct physics. Note that fermion doubling appears for each dimension that is discretized, *i.e.* in d dimensions there are 2^d solutions for every physical solution.

Comparing the eigenfunctions for $k = 0$ and $k = \pi/a$ of the discretized Hamiltonian gives a tangible explanation for the appearance of the doubled modes. The mode with $k = 0$ is constant in space, thus the spatial derivative vanishes. For $k = \pi/a$, the plane wave alternates between plus and minus one on neighboring sites [see Fig. 2.4 c)]. The spatial derivative at point j is given by $(\Psi_{j+1} - \Psi_{j-1})/2a$, which vanishes since Ψ_{j+1} and Ψ_{j-1} have the same value. This means that the central difference quotient does not capture the highly oscillating wave function, which is at the heart of the problem. However, the central difference quotient is crucial if Hermitian Hamiltonian matrices are required. The problem with fermion doubling is that the spurious modes can affect the simulated observable, yielding incorrect results. Fortunately, there are several ways to circumvent fermion doubling [91], each of which has its own advantages and disadvantages. In this thesis, we use two of these approaches, and they are presented in the following.

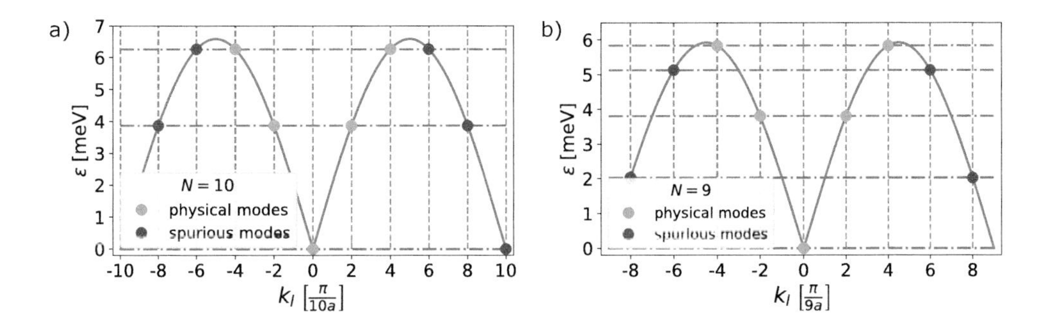

Figure 2.5: The blue line shows the dispersion given by Eq. (2.22) for $\epsilon > 0$ and a) $N = 10$ and b) $N = 9$ lattice points. The length of the system is $L = 500$ nm. The gray vertical lines mark the allowed k_l-values and the purple horizontal lines give the eigenenergies of the system. For an even number of lattice points the physical and the spurious solutions are degenerate, whereas for an odd number they are non-degenerate.

The first method relies on the fact that the spurious solutions only affect the results in a non-trivial way if they couple to the physical solutions. The physical and the doubled Dirac cone can be regarded as two distinct valleys in the Brillouin zone and coupling occurs only via inter-valley scattering. Thus, for clean systems, each valley lives on its own and provides the correct physics. There are, however, additional degeneracies (depending on the number of dimensions where fermion doubling occurs) which have to be taken into account. For instance, when computing the conductance with absent inter-valley scattering, the correct result is obtained by simply dividing by the number of fermion doubling induced degeneracies 2^d.

Note that there appears a difficulty when dealing with periodic systems with finite size. The spurious solutions might not be degenerate with the physical ones, making it difficult to remove the effect of fermion doubling. To elaborate on that, let us return to the 1D toy model with finite length L and a finite number of lattice points N. With $L = Na$ the allowed wavenumbers can be written as $k_l = \frac{2l}{N}\frac{\pi}{a}$, where l is an integer satisfying $-N/2 < l \leq N/2$. Consider, for the moment, a wavenumber k_l which corresponds to a physical mode, and, without loss of generality, $k_l < 0$. There exists only a spurious mode which is degenerate with the physical one if the wavenumber $k_{l'} = k_l + \pi/a$ exists, since $|\sin(k_l + \pi/a)\, a| = |\sin(k_l a)|$. A shift of k_l by π/a corresponds to a shift of l by $N/2$. However, l' must be an integer, thus $l' = l + N/2$ only exists if N is an even number. Hence, physical and spurious modes are only degenerate if an even number of lattice points is used. As an example, let us compare the cases with $N = 10$ and $N = 9$ which are depicted in Fig. 2.5. For $N = 9$, the spurious and physical states are not degenerate, and thus method one to avoid fermion doubling by simply removing the effect of the additional degeneracies does not

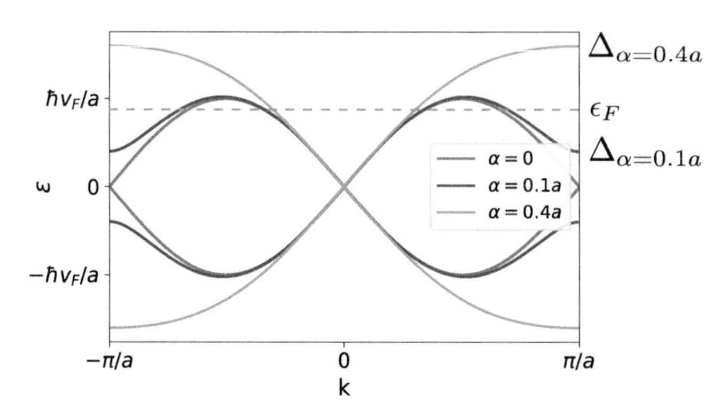

Figure 2.6: Spurious solutions are removed in an energy window $\pm\Delta = \pm 4v_F\hbar\alpha/a^2$ by a Wilson mass term. While the doublers still exist at ϵ_F (gray line) for a value of $\alpha = 0.1a$, they are absent for $\alpha = 0.4a$.

work. Note that an additional Aharonov-Bohm type phase (*e.g.* due to a magnetic field), which leads to a shifted wavenumber quantization $k_l \to k_l + \Delta k$, does not spoil the argument.

Another difficulty appears when treating disordered systems since coupling between physical and doubled modes needs to be suppressed. White noise disorder, also called Anderson disorder, where a random onsite energy is assigned to each lattice site (without any correlation), cannot be used since it allows scattering with a large momentum transfer, which in turn facilitates inter-valley scattering. Inter-valley scattering can be avoided by using smooth disorder, *i.e.* correlated disorder where the correlation length ξ is large compared to the lattice spacing a.

The second method is the so-called Wilson's mass approach, which was recently discussed in the context of quantum transport through nanojunctions in Ref. [92]. The idea is to add a small mass term quadratic in the momentum, which results in a Hamiltonian of the form

$$\hat{H} = \hbar v_F \left(\hat{k}_x\sigma_x + \alpha\hat{k}_x^2\sigma_y\right), \tag{2.23}$$

where the strength of the mass term is tuned by the parameter α. Note that any linear combination of σ_y and σ_z can be used to open the mass gap. Discretizing the Dirac equation for the above Hamiltonian by using Eq. (2.14) and Eq. (A.1) yields the set of equations

$$\hbar v_F\left[-\mathrm{i}\sigma_x\frac{1}{2a}\left(\Psi_{j+1}-\Psi_{j-1}\right)-\alpha\sigma_y\frac{1}{a^2}\left(\Psi_{j+1}-2\Psi_j+\Psi_{j-1}\right)\right]=\epsilon_\alpha\Psi_j. \tag{2.24}$$

This set of equations is solved by the ansatz $\Psi_j = \frac{1}{\sqrt{L}}\mathrm{e}^{\mathrm{i}kja}\chi$, which yields the dispersion

$$\epsilon_\alpha(k_x) = \pm\hbar v_F\sqrt{\frac{1}{a^2}\sin^2(k_xa)+\alpha^2\frac{4}{a^2}\left[1-\cos(k_xa)\right]^2}. \tag{2.25}$$

The mass term opens a gap $\Delta = 4v_F\hbar\alpha/a^2$ at the border of the Brillouin zone ($k_x = \pi/a$) and thus can be used to remove the spurious modes from the desired energy range, while the massless spectrum at $k_x = 0$ stays intact. This can be seen in Fig. 2.6 which shows $\epsilon_\alpha(k_x)$ for several values of α. The drawback of this method is that it breaks TRS and modifies the spin texture which can have an effect on the simulated observables. Thus, the Wilson's mass approach has to be used with care.

Throughout this thesis, we resort to method one whenever possible. If method one is not applicable, we resort to the Wilson's mass approach and carefully check the results for any unwanted effects. We will give further details about Fermion doubling in cylindrical 3DTI nanowires in Secs. 3.1.2 and 3.2, and in 3DTI nanocones in Sec. 5.3.

3

Topological insulator nanowires with constant cross section

In this chapter, we use the basic theoretical and numerical concepts that were introduced in Ch. 2 and apply them to 3DTI nanowires with constant cross section. Such nanowires consist of a 3DTI material and are surrounded by a trivial insulator. According to the bulk-boundary correspondence (see Sec. 2.3.2), TSS, which are in the simplest case described by a 2D Dirac Hamiltonian (see Sec. 2.3.1), form at the interface between both materials.

Such nanowires were studied experimentally for instance in Refs. [17, 19–24], and theoretically in Refs. [18, 25–27]. To the best of our knowledge, transport through these wires has not been studied using a pure surface theory in combination with a tight-binding wave function matching approach. Either a full bulk description, or the transfer matrix method [93–95] has been used so far. Hence, we study transport through nanowires with constant cross section in much detail in this chapter, which will help us to familiarize with the peculiarities of the methods we use. In particular, we will learn how to set up the correct tight-binding Hamiltonian for Dirac electrons on a 2D cylindrical surface, and we will discuss the effect of fermion doubling in such systems. Also, we will explain how to effectively implement disorder, and how to model realistic contacts. All of this is crucial when proceeding to more complicated systems discussed in later chapters. Moreover, it is worthwhile to discuss the physics of nanowires with constant cross section in detail, which will be important especially in Ch. 4.

We use a pure surface theory to describe transport, which means that we neglect all bulk contributions. This is reasonable since we are only interested in the characteristic transport features of the TSS. Note that the shape of the cross section of the nanowires used in experiments is often rectangular. However, as long as the surface states can be described by a Hamiltonian linear in the momentum, the effective surface Hamiltonian is equivalent to that of a cylinder [96, 97].[1]

[1] This is not true as soon as the cross section changes along the wire direction, which we will elaborate on in Sec. 6.3.

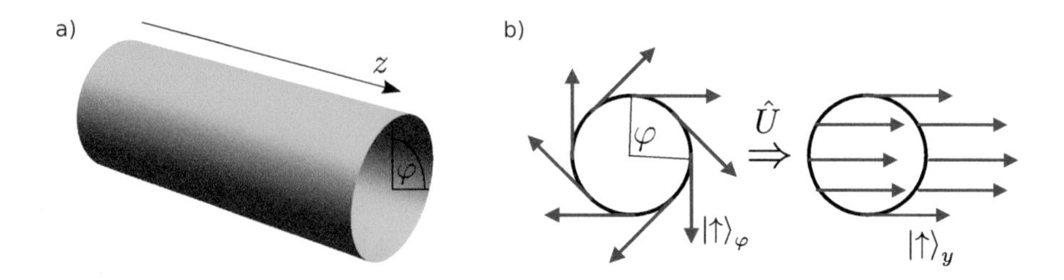

Figure 3.1: a) Sketch of a cylindrical 3DTI nanowire surface. b) Spin texture around the circumference of the nanowire for clockwise angular motion. Before the local rotation with $\hat{U}$, the spin follows the momentum, and hence winds around the circumference. After applying $\hat{U}$, its component in the transversal direction always points in the y-direction.

Although we discuss cylindrically shaped nanowires in the following, the results are thus also valid for nanowires with cross sections different from a circle.

3.1 Hamiltonian of the clean nanowire

Let us start with an infinitely long cylindrical 3DTI nanowire, whose surface is sketched in Fig. 3.1 a). The Hamiltonian describing the Dirac-like states living on the cylindrical surface is based on the flat space Hamiltonian for Dirac electrons $\hat{H}_{\text{flat}} = \hbar v_F \left(\hat{k}_x \sigma_x + \hat{k}_y \sigma_y\right)$ introduced in Sec. 2.3.1. From now on, we will omit the "hat" symbol on top of Hamiltonians for convenience. Changing to a cylindrical geometry modifies H_{flat}, which gives

$$H = \hbar v_F \left[\hat{k}_z \sigma_z + \frac{1}{2}\left(\hat{k}_\varphi \sigma_\varphi + \sigma_\varphi \hat{k}_\varphi\right)\right], \tag{3.1}$$

where we use standard cylindrical coordinates (z, φ), the Pauli matrix $\sigma_\varphi \equiv \sigma_y \cos\varphi - \sigma_x \sin\varphi$, and the angular momentum wavenumber operator $\hat{k}_\varphi = -i\frac{1}{R}\partial_\varphi$ with radius R. The Hamiltonian H is obtained by simply substituting Cartesian with cylindrical coordinates in H_{flat}, *i.e.* $x \to z$ and $y \to \varphi$. The additional symmetrization between Pauli matrix σ_φ and wavenumber operator $\hat{k}_\varphi$ is necessary because the term $\sigma_\varphi \hat{k}_\varphi$ alone is not Hermitian (since σ_φ and $\hat{k}_\varphi$ do not commute). Eq. (3.1) can also be rigorously derived by including the so-called *spin connection* [25, 96, 98], which will be discussed briefly in the context of curved/shaped 3DTI nanowires in Sec. 6.1.

We can simplify Eq. (3.1) by applying the local unitary transformation $\hat{U}(\varphi) = \exp(-i\varphi\sigma_z/2)$, which corresponds to a rotation of the spin around the z-axis by an angle of φ in the Bloch sphere [see Fig. 3.1 b)]. The transformed Hamiltonian $\tilde{H} \equiv \hat{U}^{-1} H \hat{U}$ is derived in App. A.2 and reads

$$\tilde{H} = \hbar v_F \left(\hat{k}_z \sigma_z + \hat{k}_\varphi \sigma_y \right). \tag{3.2}$$

Interestingly, the only difference to H_{flat} is that $\hat{k}_y$ is replaced by $\hat{k}_\varphi$, which means that $\tilde{H}$ is flat in spinor space. However, the transformation with $\hat{U}^{-1}(\varphi)$ modifies the boundary condition, since

$$\begin{aligned} \hat{U}^{-1}(\varphi + 2\pi) &= \mathrm{e}^{i\varphi\sigma_z/2} \mathrm{e}^{i\pi\sigma_z} \\ &= U(\varphi) \sum_{n=0}^{\infty} \frac{(\mathrm{i}\pi\sigma_z)^n}{n!} \\ &= U(\varphi) \left[\sigma_0 \left(1 - \frac{\pi^2}{2} + \frac{\pi^4}{4!} - \ldots \right) + \mathrm{i}\sigma_z \left(\pi - \frac{\pi^3}{3!} + \frac{\pi^5}{5!} - \ldots \right) \right] \\ &= U(\varphi) \left[\sigma_0 \cos\pi + \mathrm{i}\sigma_z \sin\pi \right] \\ &= -U(\varphi), \end{aligned} \tag{3.3}$$

where we grouped the terms of the Taylor expansion of the exponential in line 3 such that Taylor expansions of cosine and sine function appear in line 4. Here, σ_0 denotes the identity matrix in two dimensions. It follows that the transformed wave functions satisfy antiperiodic boundary conditions, since

$$\tilde{\Psi}(z, \varphi + 2\pi) = U^{-1}(\varphi + 2\pi)\Psi(z, \varphi + 2\pi) = -U^{-1}(\varphi)\Psi(z, \varphi) = -\tilde{\Psi}(z, \varphi), \tag{3.4}$$

where we used that the original wave functions satisfy periodic boundary conditions. Notably, the minus sign can be viewed as a curvature-induced Berry phase [18][2] which accounts for the fact that if the electron travels around the circumference its spin is rotated by 2π (due to spin-momentum locking), and hence the spinor acquires a phase of π.

3.1.1 Analytical solution

The Hamiltonian $\tilde{H}$ given by Eq. (3.2) is the starting point for our analysis of the transport properties of cylindrical 3DTI nanowires. In this section, we derive the spectrum of $\tilde{H}$, and discuss the consequences of the antiperiodic boundary conditions. Due to rotational symmetry, we can use the ansatz $\exp(\mathrm{i}k_\varphi R\varphi)$ for the transversal part of the eigenfunction $\tilde{\Psi}(z, \varphi)$ of $\tilde{H}$. The antiperiodic boundary conditions derived in Eq. (3.4) imply

$$\exp[\mathrm{i}k_\varphi R(\varphi + 2\pi)] = -\exp(\mathrm{i}k_\varphi R), \tag{3.5}$$

[2] Sometimes the curvature-induced Berry phase is referred to as *spin* Berry phase, as for instance in Ref. [99].

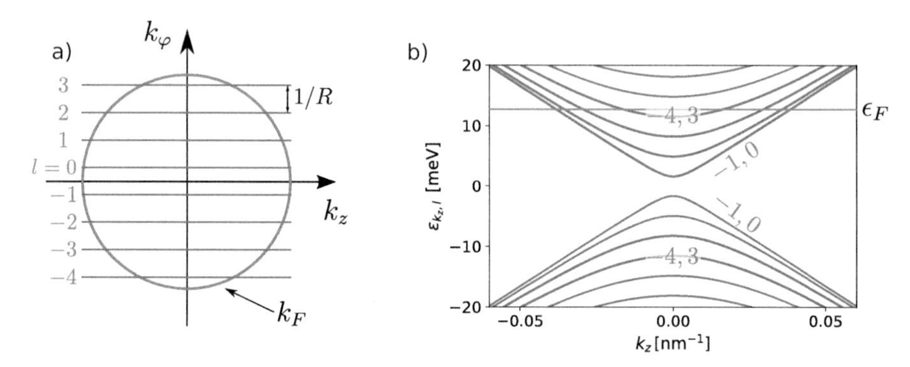

Figure 3.2: a) The angular wavenumber k_φ takes discrete values $(l+1/2)/R$ while k_z is continuous (blue lines). Each line is labeled by its angular momentum quantum number l. The green circle depicts the Fermi surface of a 2D Dirac cone in flat space. Cuts of this Dirac cone at the allowed angular wavenumbers yield the subband structure in b). b) Dispersion of a cylindrical 3DTI nanowire with radius $R = 100\,\mathrm{nm}$ and Fermi velocity $v_F = 5 \times 10^5\,\mathrm{m\,s^{-1}}$ (we will use this value for v_F throughout the dissertation). Each subband is two-fold degenerate, and a few of the subbands are labeled with their corresponding angular momentum quantum number l. The Fermi energy which yields the Fermi surface in a) is marked with a horizontal green line.

which leads to an angular wavenumber quantization of $k_\varphi = (l+1/2)/R$, with $l \in \mathbb{Z}$. Note that the usual finite size quantization of the angular wavenumber due to the finite circumference of the nanowire is shifted by 1/2 as a result of the non-trivial Berry phase. Figure 3.2 a) depicts the angular wavenumber quantization together with the Fermi surface of a 2D Dirac cone (green). Each blue horizontal line corresponds to one angular momentum quantum number l and a continuous k_z. Note that there is no line for $k_\varphi = 0$ because $l + 1/2 \neq 0$ for all l.

The spectrum of the cylindrical nanowire is readily obtained using the ansatz $\tilde{\Psi}(z,\varphi) \propto \exp(\mathrm{i}k_\varphi R\varphi)\exp(\mathrm{i}k_z z)\,\tilde{\chi}$ (where $\tilde{\chi}$ is a two-component spinor), and reads

$$\epsilon_{k_z,l} = \pm\hbar v_F \sqrt{k_z^2 + \frac{1}{R^2}\left(l+\frac{1}{2}\right)^2}, \tag{3.6}$$

which is plotted in Fig. 3.2 b). The spectrum $\epsilon_{k_z,l}$ is the same as in flat space given by Eq. (2.6), except for the quantization of the angular wavenumber. Each subband (blue line) originates from "cutting" the 2D Dirac cone associated with H_{flat} at the allowed k_φ-values [cf. Fig. 3.2 a)] and projecting the cuts into the plane. The subband is two-fold degenerate since clockwise and anticlockwise moving partners with the same $|k_\varphi|$-value have the same energy. Note that due to the shifted angular wavenumber quantization $l + 1/2$ (*i.e.* due to the Berry phase) the Dirac spectrum of Fig. 3.2 b) is massive, meaning that there is a gap associated with a mass (which is the angular wavenumber). At first sight, this might seem like a contradiction to the bulk-boundary correspondence, which states that the band gap has to close at the interface between topologically trivial and non-trivial phase (see Sec. 2.3.2). However, the bulk-boundary correspondence still holds in the sense that the boundary of the 3D system (the 2D surface of the cylinder) hosts 2D massless Dirac fermions. The reason for the 1D massive Dirac spectrum is the finite circumference of the cylinder together with the curvature-induced Berry phase acquired by the spin rotation, which does not allow angular wavenumbers $k_\varphi = 0$.

The eigenspinors $\tilde{\chi}$ of the cylindrical nanowire have the same structure as the eigenspinors χ_{flat} in flat space given by Eq. (2.8) and (2.9). In order to obtain $\tilde{\chi}$, one only needs to substitute k_y with k_φ in χ_{flat} and rotate the spinor by $-\pi/2$ around the y-axis (since σ_z is used in $\tilde{H}$ instead of σ_x).

3.1.2 Numerical approach

In order to analyze the transport characteristics of 3DTI nanowires, we resort to numerical simulations using *kwant* (see Sec. 2.4). The Hamiltonian $\tilde{H}$, given by Eq. (3.2), is an effective continuum Hamiltonian. In order to obtain a tight-binding model that can be used by *kwant*, we discretize $\tilde{H}$ by using the finite difference method introduced in Sec. 2.4.1. This yields

$$\tilde{H}_{\text{TB}} = -\hbar v_F \frac{\mathrm{i}}{2a} \sum_{i,j} \left(\sigma_z \left| i, j \right\rangle \left\langle i+1, j \right| + \sigma_y \left| i, j \right\rangle \left\langle i, j+1 \right| \right) + \text{h.c.}, \tag{3.7}$$

where we assume equal lattice constants a in the longitudinal and transversal directions. The corresponding lattice is shown in Fig. 3.3. The real space coordinates on the lattice $(z_i, R\varphi_j)$ are connected to the indices (i, j) via $(z_i, R\varphi_j) \equiv (ia, ja)$. Due to the unitary transformation $U(\varphi) = \exp(-i\varphi\sigma_z/2)$ we applied, we need to impose anti-periodic boundary conditions. This can be implemented for instance by adding an additional minus sign in the hoppings from $(z_i, 0)$ to $(z_i, R\varphi_N)$ for all i.[3] One of those hoppings is marked by the red line in Fig. 3.3.

The band structure of the infinite cylindrical nanowire obtained with *kwant* is shown in Fig. 3.4 a). Here, the full Brillouin zone of the numerical lattice is shown. Apparently, there is a copy of the central ($k_z = 0$) Dirac cone at the border of the Brillouin zone ($k_z = \pm\pi/a$). This can be attributed to fermion doubling (see Sec. 2.4.2), which originates from the discretization of the longitudinal wavenumber operator $\hat{k}_z$. Note that there is additional fermion doubling due to the discretization

[3] Note that the phase of π can be added to any hopping in the transversal direction, or redistributed to several hoppings.

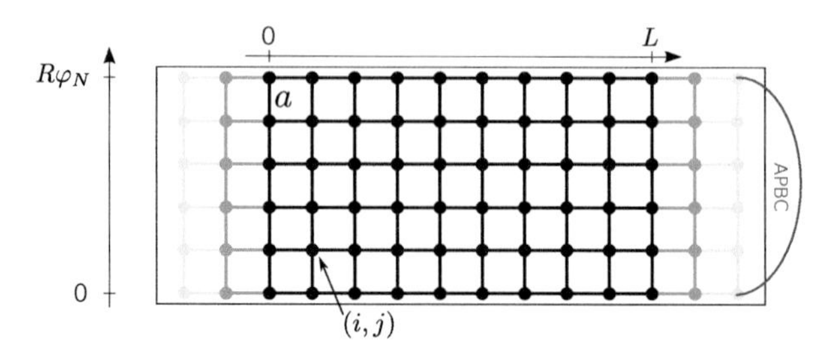

Figure 3.3: Numerical lattice created with *kwant* with $N = 6$ transversal lattice points. Each dot represents a lattice site and the lines connecting the dots represent hoppings. For a cylindrical 3DTI nanowire, there are additional hoppings in the transversal direction from 0 to $R\varphi_N$ which ensure antiperiodic boundary conditions (APBC) [only one of those hoppings is shown (red line)]. The blue dots represent semi-infinite leads, where only the first two unit-cells are shown (with fading color).

of the transversal wavenumber operator $\hat{k}_\varphi$. Since we used an even number of lattice points in the transversal direction, the physical and the spurious modes are degenerate. Hence, the bands in Fig. 3.4 a) are fourfold degenerate – a factor of 2 coming from fermion doubling due to the discretization of $\hat{k}_\varphi$, and another factor of 2 due to the degeneracy of clockwise and anticlockwise moving electrons with the same $|k_\varphi|$-value. Fig. 3.4 b) shows a zoom of the band structure depicted in a) in the low-energy regime and for small k_z. It is in good agreement with the analytical solution shown in Fig. 3.2.

In order to compute the conductance, we attach two semi-infinite leads (blue dots in Fig. 3.3) which act as wave guides. We use the same Hamiltonian for the leads as for the cylinder. From the Landauer Büttiker formalism (see Sec. 2.1), we expect that each mode at the Fermi energy contributes e^2/h to the conductance since there is nothing which causes scattering. Indeed, this is what we see if we compare Fig. 3.4 c), where the conductance as a function of the Fermi energy ϵ_F is shown, with the band structure in Fig. 3.4 b). Note that we divided the conductance obtained with *kwant* by 4 in order to account for fermion doubling. The deviation between the onset of the conductance steps and the opening of the subbands (for the analytical solution of the continuum Hamiltonian) for larger energies ($\epsilon \gtrsim 10$ meV) can be attributed to the lattice approximation, which is only valid as long as $k_\varphi a \ll 1$ (see Sec. 2.4.1).

3.2 Disordered nanowires

The samples used in experiments are never perfectly clean. There is always some degree of dirt/disorder coming from, for instance, adatoms, vacancies, or lattice

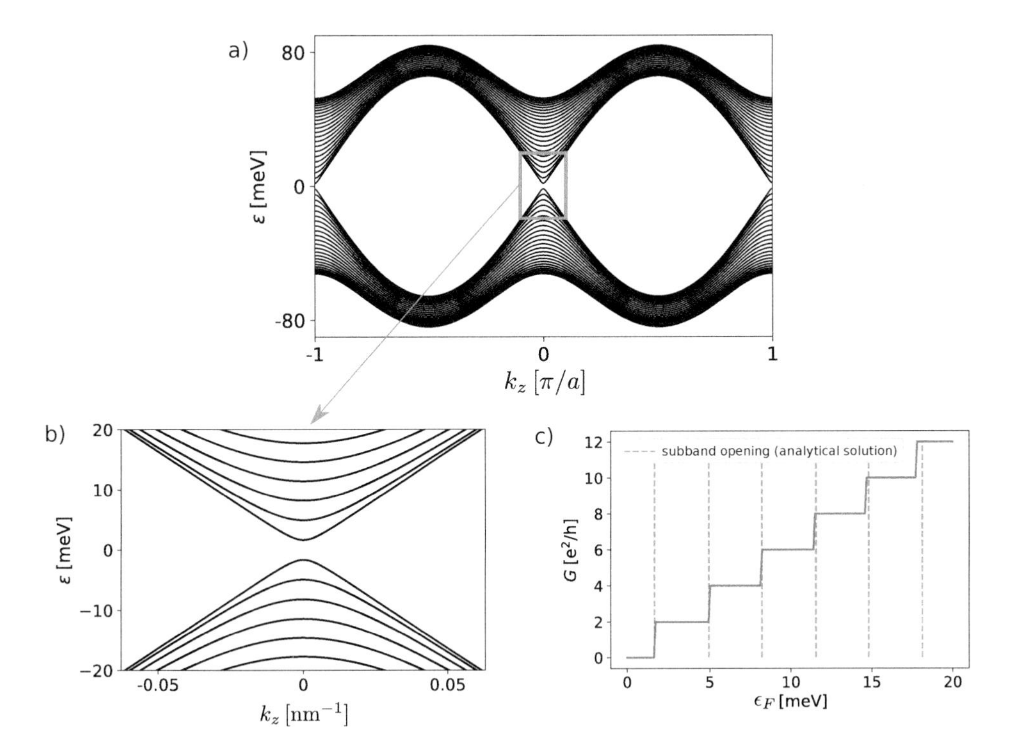

Figure 3.4: a) Band structure of the cylindrical 3DTI nanowire in the full Brillouin zone of the numerical lattice. Note how the Dirac cone reappears at the edge of the Brillouin zone, which can be attributed to fermion doubling. b) Zoom of the band structure around $k_z = 0$ [marked by the green rectangle in a)]. For low energies and small k_z-values, the correct dispersion is obtained by the tight-binding model [cf. Fig. 3.2 b)]. c) Conductance as a function of the Fermi energy together with the position of the subband openings for the analytical solution of the continuum Hamiltonian. We use a cylinder with radius $R = 100$ nm, a longitudinal lattice spacing of $a = 5$ nm, $N = 100$ lattice points in the transversal direction, and a Fermi velocity of $v_F = 5 \times 10^5\,\mathrm{m\,s^{-1}}$.

dislocations. We simulate disorder by adding a random potential $V_{\mathrm{dis}}(\boldsymbol{r})$ to the Hamiltonian. In order to extract the relevant physics, we then take an average of the observable over many different random potentials (typically we use around 100 to 1000), each of which corresponds to one disorder configuration. In the following, we discuss the form of $V_{\mathrm{dis}}(\boldsymbol{r})$.

A convenient choice to simulate disorder within a tight-binding model is to add the so-called *white noise* or *Anderson* disorder. Here, for each lattice point, a random onsite energy is drawn, for instance, from a box distribution $[-W, W]$ with disorder amplitude W. Using this method has some drawbacks. First, the effective correlation length is given by the lattice spacing a, which has nothing to do with the physical system. The mean free path, however, usually depends on the correlation length of the disorder and thus on a. As a consequence, the disorder amplitude W is not meaningful on its own. One can often work around this problem by doing an effective rescaling of W to remove the lattice spacing dependence (see, for instance, Ref. [100]). The second drawback is that Anderson disorder induces inter-valley scattering and thus couples spurious and physical modes if fermion doubling is present, which can spoil the results. Therefore, method one to avoid fermion doubling (see Sec. 2.4.2) cannot be used with Anderson disorder. One has to remove the spurious solutions from the energy window by a Wilson mass (method two), which has, as discussed, its own disadvantages.

Mainly due to the second drawback, we resort to correlated disorder (unless stated otherwise) with a correlation length $\xi \gg a$. Implementing correlated disorder is more involved and will be discussed in the following.

3.2.1 Correlated disorder

We create correlated disorder by using the so-called *Fourier filtering method* (FFM), which is discussed in detail for instance in Ref. [101]. The Fourier filtering method (FFM) is extremely fast, making it possible to construct correlated disorder configurations for large systems and also in higher dimensions (we will use the FFM in 2D and 3D). Another advantage is that the resulting disorder potential has periodic boundary conditions, which is very useful when studying nanowires.

The elastic scattering time τ_k of an electron with wavenumber $k \equiv |\boldsymbol{k}|$ in a disorder potential $V_{\mathrm{dis}}(\boldsymbol{r})$ can be calculated using Fermi's golden rule, which yields (see, for instance, Ref. [102])

$$\frac{\hbar}{\tau_k} = \frac{1}{2\pi} \int \mathrm{d}^2\boldsymbol{k}'\, \mathrm{d}^2\boldsymbol{r}\; \delta(\epsilon_k - \epsilon_{k'})\, \langle V_{\mathrm{dis}}(0) V_{\mathrm{dis}}(\boldsymbol{r})\rangle\, \mathrm{e}^{\mathrm{i}(\boldsymbol{k}-\boldsymbol{k}')\cdot\boldsymbol{r}}. \tag{3.8}$$

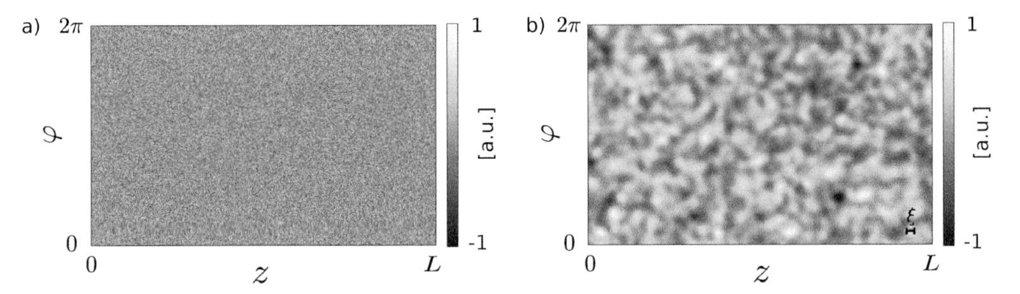

Figure 3.5: a) Anderson disorder and b) Gaussian correlated disorder. Here, we use a cylinder of radius $R = 100$ nm, length $L = 1000$ nm and a lattice with $N_\varphi = 400$ points in the transversal direction and $N_z = 636$ points in the longitudinal direction. The correlation length in b) is $\xi = 15$ nm.

The disorder potential enters the formula for the elastic scattering time via the correlator $\langle V_{\text{dis}}(0)V_{\text{dis}}(\boldsymbol{r})\rangle$, where $\langle ... \rangle$ denotes the disorder average for an infinite number of disorder realizations, *i.e.*

$$\langle V_{\text{dis}}(0)V_{\text{dis}}(\boldsymbol{r})\rangle = \lim_{N_{\text{dis}}\to\infty} \frac{1}{N_{\text{dis}}} \sum_{l=1}^{N_{\text{dis}}} V_{\text{dis}}^{(l)}(0)V_{\text{dis}}^{(l)}(\boldsymbol{r}). \tag{3.9}$$

Here, the number of disorder configurations is denoted by N_{dis} and each individual configuration is labeled by l. From Eq. (3.8) it follows that the strength of the disorder is determined by the correlator Eq. (3.9). Therefore, it makes sense to use the correlator as a definition for the type of disorder used in a simulation. For our simulations, we use a correlator which is often used in the literature (see, for instance, Refs. [18, 95, 103]) and thus allows to compare results easily. It is given by

$$\langle V_{\text{dis}}(0)V_{\text{dis}}(\boldsymbol{r})\rangle = K\frac{\hbar v_F}{2\pi\xi^2}e^{-r^2/2\xi^2}, \tag{3.10}$$

i.e. $V_{\text{dis}}(\boldsymbol{r})$ is Gaussian correlated. The amplitude K is a dimensionless measure for the disorder strength. Note that the qualitative results do not depend on the precise form of the correlator.

In order to create a suitable disorder potential for the cylindrical nanowire, we use the FFM in 2D on the lattice of the tight-binding model, which gives us an onsite potential $V_{\text{dis}}(z_i, R\varphi_i)$ for every lattice point. We then add $V_{\text{dis}}(z_i, R\varphi_i)\sigma_0$ to the tight-binding Hamiltonian $\tilde{H}_{\text{TB}}$ given by Eq. (3.7). An example of a correlated disorder landscape on a cylinder and how it compares to Anderson disorder is shown in Fig. 3.5. In Sec. 5.3, we will explain how to create correlated disorder on curved 2D surfaces.

3.2.2 Doping the leads

In transport simulations, we mimic the contacts used in experiments by attaching semi-infinite leads to the scattering region as shown in Fig. 3.3. Those leads act as wave guides leading electronic modes into and out of the scattering region. Naturally, the question arises which Hamiltonian is suitable to describe the leads.

One possibility is to use leads which are described by a semi-infinite version of the Hamiltonian of the scattering region (as we did for the clean cylinder in Sec. 3.1.2). However, this method only works if the Hamiltonian of the scattering region is translationally invariant, which is usually not the case. Disordered or shaped/curved scattering regions break the translational invariance. One way to circumvent this problem is to modify the Hamiltonian of the lead by hand by removing the disorder or adapting the shape. For instance, clean cylindrical leads can be added to any weirdly shaped and disordered nanowire.

However, in transport experiments usually metallic wide contacts with a large number of modes are attached to the system. This situation is sometimes not accurately described by the simulation if the leads host only few modes. One can solve this problem by adding a negative onsite energy ϵ_{lead} to the lead Hamiltonian, which allows to tune the number of modes by hand. The correct results are obtained when the observables are converged with respect to the number of modes. As we will see later on, there are cases where this procedure is crucial. For instance in Ch. 5, we will present a nanocone in coaxial magnetic field which hosts many degenerate QH states at zero energy which enable transport, but the cylindrical leads host zero modes.

Note that we smooth the potential steps induced by ϵ_{lead} at the interfaces between leads and system in order to suppress Fabry-Pérot interferences. Moreover, if method one to avoid Fermion doubling is used (see Sec. 2.4.2), a smooth potential step is necessary to prevent inter-valley scattering between spurious and physical solutions.

3.3 Coaxial magnetic field and the perfectly transmitted mode

Starting from Eq. (3.2) and using minimal coupling, the Hamiltonian for the cylinder in longitudinal/coaxial magnetic field $\boldsymbol{B} = B\boldsymbol{e}_z$ can be written as

$$\tilde{H} = \hbar v_F \left[\hat{k}_z \sigma_z + \left(\hat{k}_\varphi + \frac{e}{\hbar} A_\varphi \right) \sigma_y \right], \tag{3.11}$$

where e is the elementary charge. Here, we use the vector potential $\boldsymbol{A}$ in the symmetric gauge, *i.e.* $\boldsymbol{A} = BR/2\,\mathbf{e}_\varphi \equiv A_\varphi \boldsymbol{e}_\varphi$, where $\boldsymbol{e}_\varphi$ is the azimuthal unit vector. The vector potential can be removed from the Hamiltonian with the local unitary gauge transformation $U_B(\varphi) \equiv \exp\left(-\frac{\mathrm{i}}{\hbar} e A_\varphi \varphi\right)$, which yields $\bar{H} \equiv U_B^{-1}(\varphi)\tilde{H}U_B(\varphi) = \hbar v_F \left(\hat{k}_z \sigma_z + \hat{k}_\varphi \sigma_y\right)$. This modifies the boundary conditions of the transformed wave functions $\bar{\Psi} \equiv U_B^{-1}\tilde{\Psi}$, which read

$$\bar{\Psi}(\varphi + 2\pi) = -\exp(-\mathrm{i}2\pi\Phi/\Phi_0)\bar{\Psi}(\varphi), \tag{3.12}$$

where $\Phi = \pi R^2 B$ is the magnetic flux through the cross section of the cylinder, and $\Phi_0 \equiv h/e$ is the magnetic flux quantum. The factor of -1 in front of the exponential is the curvature-induced Berry phase derived in Sec. 3.1 [(see Eq. (3.4)]. Note that we can already say that the spectrum of $\bar{H}$ is Φ_0 periodic since the wave functions $\bar{\Psi}$ are Φ_0-periodic.

As usual, the Dirac equation $\bar{H}\bar{\Psi} = \epsilon\bar{\Psi}$ can be solved with the Ansatz $\bar{\Psi} \propto \exp(\mathrm{i}k_z z)\exp(\mathrm{i}k_\varphi R\varphi)\bar{\chi}$, where the boundary conditions lead to the quantization of the angular wavenumber $k_\varphi = (l + 1/2 - \Phi/\Phi_0)/R$. The spectrum for a cylinder with radius R and axial magnetic field B is then given by [cf. Eq. (3.6)]

$$\epsilon_{k_z,l}(B) = \pm\hbar v_F \sqrt{k_z^2 + \frac{1}{R^2}\left(l + \frac{1}{2} - \frac{\Phi(R,B)}{\Phi_0}\right)^2}. \tag{3.13}$$

In Fig. 3.6 we plot the spectrum $\epsilon_{k_z,l}(B)$ for three distinct magnetic fluxes. Due to the periodicity of $\epsilon_{k_z,l}(B)$ in the magnetic flux with a period of one flux quantum Φ_0, the cylinder with $\Phi = n\Phi_0$, where $n \in \mathbb{Z}$, is equivalent to the cylinder with $\Phi = 0$, which we treated in Sec. 3.1.1. Thus, Fig. 3.6 a) shows the same spectrum as already presented in Fig. 3.2 b). Since states with angular momentum quantum numbers l and $-l-1$ yield the same energies, each subband is doubly degenerate. However, by adding a small magnetic flux of $0.1\Phi_0$ as shown in Fig. 3.6 b), the degeneracy is lifted and bands with positive angular momentum move down while bands with negative angular momentum move up in energy. At $\Phi = (n + 0.5)\Phi_0$, which is shown in Fig. 3.6 c), the curvature-induced Berry phase, responsible for the $+1/2$ in Eq. (3.13), is canceled. Consequently, there exists a mode with angular momentum quantum number l_0 such that $k_{\varphi,l_0} = 0$, which leads to the dispersion $\epsilon_{k_z,l_0}(R,B) = \pm\hbar v_F|k_z|$. Thus, the gap is closed by a 1D massless (*i.e.* linear in the momentum) Dirac mode. In the following, we discuss the properties of this massless Dirac mode, and point out its significance.

To this end, consider the two counterpropagating states $\Psi_{k_z l_0}$ and $\Psi_{-k_z l_0}$ marked with green dots in Fig. 3.13 c). Due to spin-momentum locking, we can directly locate those states on the Bloch-sphere, which is illustrated with green arrows in Fig. 3.6 d). Massless states have zero angular momentum, thus the spinor

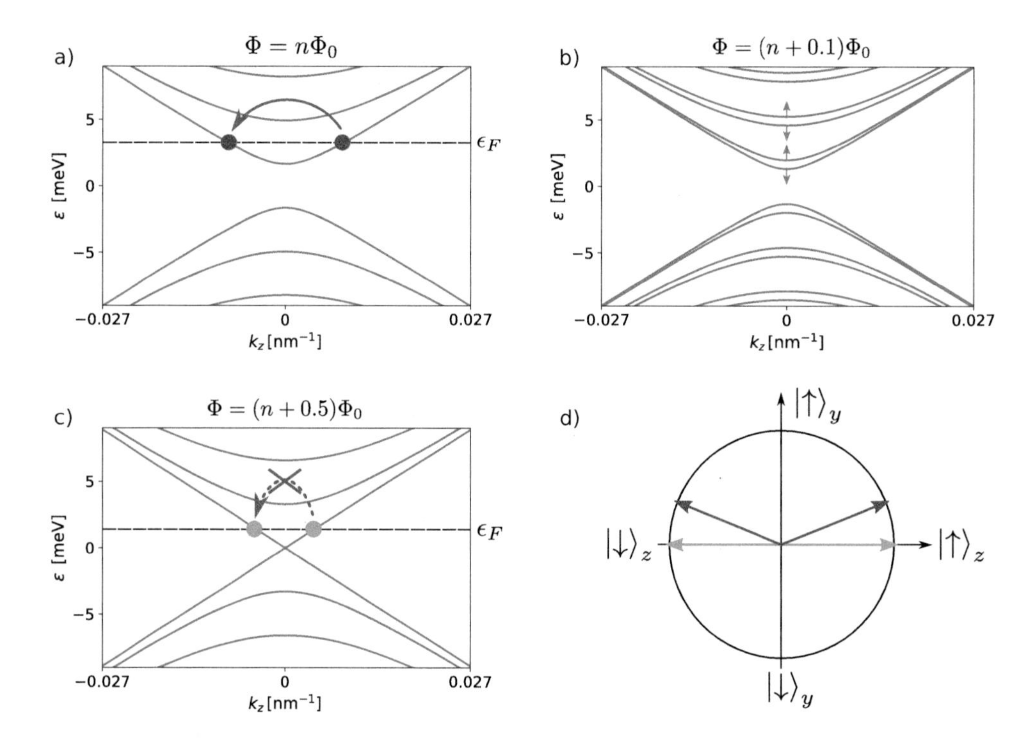

Figure 3.6: Disperson of the electronic modes of cylindrical 3DTI nanowires in coaxial magnetic field. For a magnetic flux of $\Phi = n\Phi_0$ as shown in a), the modes are two-fold degenerate. Upon slightly increasing the magnetic flux, the degeneracy splits, as shown in b) for $\Phi = (n + 0.1)\Phi_0$. The shifting of the modes with increasing flux is indicated by gray arrows. For a magnetic flux $\Phi = (n + 0.5)\Phi_0$, the gap is closed by a 1D massless mode. Panel d) shows to spin texture of the states at the Fermi energy marked in a) and c) in the y-z cross section of the Bloch sphere. The spinors of the states at the Fermi energy in a) (red dots) have a finite overlap, hence scattering between both is allowed [indicated by the gray arrow in a)]. In contrast, the two states within the massless mode shown in c) (green dots) have orthogonal spinors, and thus the scattering amplitude vanishes.

parts of the wave functions are given by $|\uparrow\rangle_z$ and $|\downarrow\rangle_z$, respectively. Due to the orthogonality of those spinors, backscattering is forbidden. Hence, we suspect perfect transmission (*i.e.* $G = e^2/h$) for $\Phi = (n + 0.5)\Phi_0$ at least as long as the Fermi level is below the first massive subband. Actually, it can be shown that if TRS is preserved and if the number of modes is odd, the conductance is at least e^2/h (see Refs. [10, 104]), independent of the Fermi level. Note that time-reversal symmetry is effectively restored, both at $\Phi = n\Phi_0$ and $\Phi = (n + 0.5)\Phi_0$, and that the number of modes is always even at $\Phi = n\Phi_0$ and always odd at $\Phi = (n+0.5)\Phi_0$. This means that the conductance is always larger than e^2/h if the flux is an odd multiple of $\Phi_0/2$, which leads to the notion of the so-called *perfectly transmitted mode*.

Now, consider the two massive states at $\Phi = n\Phi_0$ marked with red dots in Fig. 3.6 a). The bands are two-fold degenerate with respect to positive and negative orbital angular momentum. The spinors of the states with positive orbital angular momentum are illustrated on the Bloch sphere with red arrows in Fig. 3.6 d). Since they have a finite overlap, scattering between these states is allowed and the conductance is susceptible to disorder. In general, this is the case for all magnetic fluxes $\Phi \neq (n + 0.5)\Phi_0$.

Conductance simulations

In order to test our predictions from above, we resort to a numerical simulation, and compute the conductance as a function of the Fermi energy. To this end, we use the tight-binding Hamiltonian given by Eq. (3.7) to which we add a hopping that accounts for the anti-periodic boundary conditions (due to the π Berry phase). Furthermore, we add correlated disorder (see Sec. 3.2.1) to account for impurities, and highly-doped leads (see Sec. 3.2.2) to simulate contacts with many modes. Finally, we implement the magnetic field using Peierls substitution, the standard procedure to implement magnetic fields in tight-binding models [105]. Using Peierls substitution, the hopping amplitude along a path γ is modified by a phase factor $\exp\left[-\mathrm{i}\frac{e}{\hbar}\int_\gamma \mathrm{d}\boldsymbol{l}\,\boldsymbol{A}(l)\right]$.[4] Consequently, for a cylinder in coaxial magnetic field, a phase factor $\exp\left(-\mathrm{i}\frac{e}{\hbar}A_\varphi a\right)$ needs to be added in the transversal hoppings in

[4]Peierls substitution leads to the correct result if the magnetic field varies slowly on the scale of the lattice spacing. In the following, we will use a simple example to clarify the meaning of Peierls substitution. Consider a hopping from site i to $i+1$ with hopping amplitude $t = 1$. Without magnetic field, the momentum $\hat{p}$ is the generator of translations and we can write $|i+1\rangle\langle i| = \exp(\mathrm{i}\frac{e}{\hbar}\hat{p}a)|i\rangle\langle i|$. Turning on a magnetic field, the generator of translations is given by the canonical momentum $\hat{p} + e\hat{A}$. Hence, the correct hopping term is given by $\exp\left[-\mathrm{i}(p + eA)a/\hbar\right]|i\rangle\langle i|$. This can be written as $\exp\left(-\mathrm{i}\frac{e}{\hbar}Aa\right)|i+1\rangle\langle i|$, which explains the additional phase factor in the hopping. For a derivation of Peierls substitution based on Wannier functions, we refer to Ref. [105].

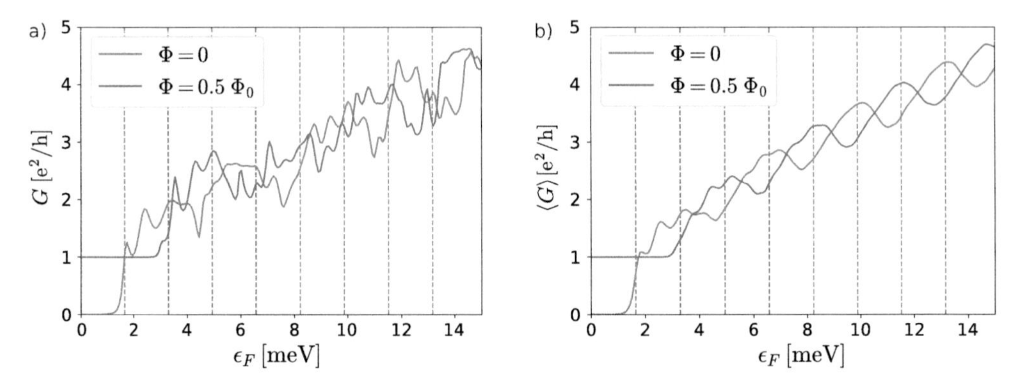

Figure 3.7: Conductance G as a function of the Fermi energy for a) one disorder configuration, and b) a disorder average. Subband openings are marked with vertical lines. For $\Phi = 0.5\Phi_0$ and below the first opening of a massive subband (first vertical blue line), a conductance plateau appears signaling perfect transmission. Here, we use a disorder strength of $K = 0.2$ with correlation length $\xi = 10\,\text{nm}$, a radius of $R = 100$ nm, and a wire length of $L = 1000$ nm. Moreover, we add a Wilson mass term in order to circumvent Fermion doubling.

Eq. (3.7). Note that this automatically leads to the correct boundary conditions of $\bar{\Psi}$ [see Eq.(3.12)].

The result of the simulation, *i.e.* the conductance as a function of the Fermi energy $G(\epsilon_F)$, is shown in Fig. 3.7 a) for one specific disorder configuration. Figure 3.7 b) shows the disorder-averaged conductance $\langle G(\epsilon_F)\rangle \equiv \frac{1}{N}\sum_i^N G_i(\epsilon_F)$, where the index i runs over $N = 300$ disorder configurations. The orange curves correspond to a magnetic flux of $\Phi = 0$, and the blue curves correspond to $\Phi = 0.5\Phi_0$. Note that $\Phi = 0$ is equivalent to $\Phi = n\Phi_0$, and $\Phi = 0.5\Phi_0$ is equivalent to $\Phi = (n + 0.5)\Phi_0$. The energies at which subbands "open" (*i.e.* are available for transport) are marked with dashed vertical lines, where the color gives the corresponding flux value.

Let us now focus on the disorder-averaged conductance $\langle G(\epsilon_F)\rangle$ shown in Fig. 3.7 b). For energies larger than ≈ 3 meV, we see an increasing conductance background with an oscillatory behavior on top, where the oscillation period has the size of the subband spacing. Note that we do not talk about the oscillations with smaller amplitude for energies $\lesssim 8$ meV, with a period smaller than the subband spacing, which stem from residual Fabry-Pérot oscillations. Those oscillations are induced by the highly doped leads and do not have any physical significance. The oscillatory behavior with a period of the subband spacing can be explained by the band structure [see Fig. 3.6 a) and c)]: Whenever the Fermi energy approaches the bottom of one of the disorder-broadened subbands, the high density of states (associated with a van Hove singularity) causes enhanced scattering[5] and thus leads to a reduction of the conductance. By further increasing the Fermi energy, the additional conductance channel fully opens and the Fermi energy leaves the vicinity of the van Hove singularity leading to an increasing conductance. The conductance can increase by more than e^2/h per subband spacing, because the massive bands are doubly degenerate for magnetic fluxes $\Phi = 0$ and $\Phi = 0.5\Phi_0$.

Another distinctive transport signature is the fact that the blue and the orange curve show an anticorrelated behavior, which can also be explained by the band structure. The position of the subband minima for $\Phi = 0.5\Phi_0$ is directly in between the position of the minima for $\Phi = 0$ [can be directly seen when setting $k_z = 0$ in Eq. (3.13)], which explains that a minimum in the conductance roughly becomes a maximum when adding half of a flux quantum.

For energies smaller than ≈ 3 meV, the conductance for $\Phi = 0$ drops to zero below the lowest subband. Note that the drop is not sudden due to disorder broadening and due the highly-doped leads. For $\Phi = 0.5\Phi_0$, however, something remarkable happens. The conductance, clearly affected by the disorder for larger energies [cf. Fig. 3.4 c)], is locked exactly at one conductance quantum $G_0 \equiv e^2/h$ below the first massive subband.

Hence, our analytical prediction turns out to be correct: the massless states are protected in the sense that they cannot backscatter. Moreover, the van Hove singularity of the first massive subband does not affect the conductance, which can be explained by the existence of the perfectly transmitted mode. For odd multiples of $\Phi_0/2$, where time-reversal symmetry is preserved and there is an odd number of modes, the conductance cannot drop below G_0 (see above). This also holds for a single disorder configuration [see Fig. 3.7 a)], and is thus not caused by an averaging mechanism.

Note that the error on the mean in Fig. 3.7 b) given by

$$\sigma(\epsilon_F) = \left[\frac{1}{N(N-1)}\sum_{i=1}^{N}(G_i(\epsilon_F) - \langle G(\epsilon_F)\rangle)^2\right]^{1/2}, \tag{3.14}$$

is smaller than $10^{-3}G_0$ for all energies and thus not shown in the plot.

We want to conclude this section by emphasizing that 3DTI nanowires are very useful to measure the robustness of topological surface states by means of transport experiments. By simply changing the flux from $\Phi = 0$ to $\Phi = \Phi_0/2$, one can induce the perfectly transmitted mode and directly proof the non-trivial nature of the TSS by measuring a conductance pinned to G_0. In experiments, however,

[5] This can be explained with Fermi's golden rule: $1/\tau \propto \rho(\epsilon_F)$, where $1/\tau$ is the inverse scattering time and $\rho(\epsilon_F)$ is the density of states at the Fermi energy.

the situation is most of the time not that simple. There might be residual bulk contribution to the conductance [15], or it might not be possible to tune the Fermi energy close enough to the Dirac point (below the first massive subband). This is the case in strained HgTe, where the Dirac point is submerged in the valence bands (see Sec. 2.3.3).

3.4 Perpendicular magnetic field – a higher-order topological insulator

By rotating the magnetic field by 90° such that it is perpendicular to the nanowire axis [see Fig. 3.9 b)], we enter a topologically distinct transport regime. We will see that if the magnetic field is strong enough such that the magnetic length $l_B = \sqrt{\hbar/eB}$ is small compared to πR, where R is the radius of the nanowire, the top and bottom surface become gapped and transport is dominated by chiral side surface states. One can argue that those chiral side surface states originate from a sign change of the magnetic field component piercing the surface with respect to the surface normal at the sides of the nanowire. This makes the nanowire in perpendicular magnetic field an extrinsic second order TI. Reviews about higher-order TIs can be found, for instance, in Refs. [29–31, 106]. TI slabs/nanoribbons in perpendicular magnetic field have been studied in [107–111]. Although the qualitative transport characteristics of TI slabs and cylinders in perpendicular magnetic field are similar, we first study the cylindrical case in detail, since it is the ideal candidate to introduce Dirac Landau levels (LLs) in curved space and the notion of an *effective mass potential.* Both concepts will be of importance in Chs. 5 and 6.

Although the TSS are no longer protected from backscattering as soon as TRS is broken by a magnetic field, we assume that their very existence is guaranteed by the band inversion argument introduced in Sec. 2.3.2. Hence, we describe the electronic surface states on the cylinder by the Dirac Hamiltonian given in Eq. 3.2, and add the perpendicular magnetic field via minimal coupling.

Dirac electrons in flat space in perpendicular magnetic field

Before we tackle the cylindrical geometry, it is instructive to consider Dirac electrons in an infinite 2D plane described by $H_{\text{flat}} = \hbar v_F(\hat{k}_x\sigma_x + \hat{k}_y\sigma_y)$.[6] For a

[6]This system is actually not physical because in reality, the surface must be closed (experimental samples are finite). However, 2D Dirac electrons subject to a perpendicular magnetic field play a central role in this thesis and we will see that it is instructive to study the infinite plane first.

magnetic field $\boldsymbol{B} = (0, 0, B)^T$ in the infinite xy-plane, the natural choice of the vector potential is the symmetric gauge $\boldsymbol{A} = B/2(-y, x, 0)^T$. However, we will apply our findings to an infinitely long cylindrical nanowire later on. In order to respect its translational invariance, we choose the Landau gauge $\boldsymbol{A} = B(0, x, 0)^T$. The Hamiltonian in real space is then given by

$$H = \hbar v_F \left[(-\mathrm{i}\partial_x)\sigma_x + \left(-\mathrm{i}\partial_y + \frac{eB}{\hbar} x \right) \sigma_y \right], \tag{3.15}$$

which is translationally invariant in the y-direction. To solve the Dirac equation, we thus use the ansatz

$$\Psi_{kn}(x, y) = \mathrm{e}^{\mathrm{i}ky} \chi_{kn}(x), \tag{3.16}$$

where $\chi_{kn}(x) \equiv (f_{kn}(x), g_{kn}(x))^T$ is a two-component spinor. The meaning of the index n will be clarified shortly. Applying H to Ψ_k, we get an effective 1D Hamiltonian for every value of k, which reads

$$H_{1D}(k) = \hbar v_F (-\mathrm{i}\partial_x)\sigma_x + \hbar v_F \left(k + \frac{eB}{\hbar} x \right) \sigma_y, \tag{3.17}$$

and defines the 1D Dirac equation $H_{1D}(k)\chi_{kn}(x) = \epsilon_n(k)\chi_{kn}(x)$. For fixed k, $H_{1D}(k)$ is comprised of a kinetic term $\hbar v_F(-\mathrm{i}\partial_x)$ and a position dependent term $V_k(x) \equiv \hbar v_F(k + \frac{eB}{\hbar}x)$, where $|V_k(x)|$ can be interpreted as an *effective mass potential* potential. The reasoning here is that the Hamiltonian (3.17) describes a Dirac electron in 1D which feels a mass term $|V_k(x)|$, in the sense that $V_k(x)$ comes with the σ_y-matrix and not with an identity matrix (which would correspond to an electrostatic potential). Hence, the electron effectively feels the absolute value of $V_k(x)$ as a potential. We can rewrite $|V_k(x)|$ as

$$|V_k(x)| = v_F e \, |B(x - x_k)| \,, \tag{3.18}$$

where $x_k \equiv -\hbar k/(eB) = -k l_B^2$. From the above equation, we can deduce that the electrons are confined in the x-direction around x_k, which is schematically depicted in Fig. 3.8 for three distinct wavenumbers. For each k, electrons feel a different effective potential, and the spatial part of $\chi_{kn}(x)$ corresponds to a series of bound states labeled by n, whose energies are sketched by the horizontal dashed lines. We will see that those bound states form LLs and can thus be associated with QH states. In the y-direction, states are still plane waves due to the translational invariance.

Solving the Dirac equation with $H_{1D}(k)$ amounts to solving the two coupled first order differential equations

$$\hbar v_F \left[-\mathrm{i}\partial_x - \mathrm{i} \left(k + \frac{eB}{\hbar} x \right) \right] g_{kn}(x) = \epsilon_n(k) f_{kn}(x) \tag{3.19}$$

$$\hbar v_F \left[-\mathrm{i}\partial_x + \mathrm{i} \left(k + \frac{eB}{\hbar} x \right) \right] f_{kn}(x) = \epsilon_n(k) g_{kn}(x), \tag{3.20}$$

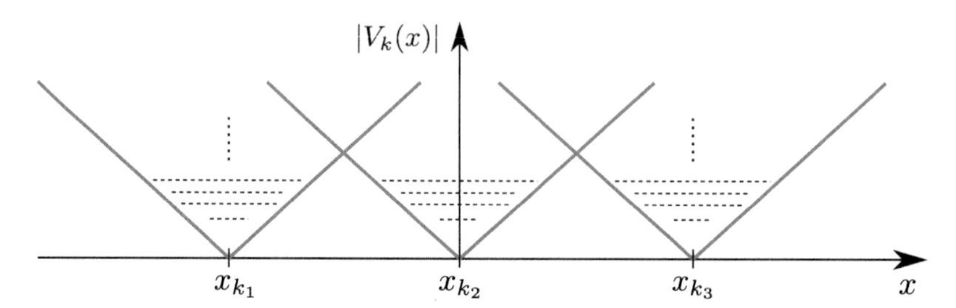

Figure 3.8: Effective potential $|V_k(x)|$ for three wavenumbers k_1, k_2, and k_3. The energies of the bound states which we identify with Dirac QH states are sketched with horizontal dashed lines.

which can be decoupled by converting them into the second order differential equation

$$\hbar^2 v_F^2 \left[-\partial_x^2 + \frac{eB}{\hbar} + \left(\frac{eB}{\hbar} x + k \right)^2 \right] f_{kn}(x) = \epsilon_n^2(k) f_{kn}(x). \tag{3.21}$$

The above equation can be brought into the form of the Schrödinger equation of an harmonic oscillator with mass m, which reads

$$\left[-\frac{\hbar^2}{2m} \partial_x^2 + \frac{1}{2} m \omega_c^2 (x - x_k)^2 \right] f_{kn}(x) = \frac{\hbar^2}{2m} \left(\frac{\epsilon_n^2(k)}{\hbar^2 v_F^2} - \frac{eB}{\hbar} \right) f_{kn}(x), \tag{3.22}$$

with the cyclotron frequency $\omega_c \equiv eB/m$. Note that for every wavenumber k we get an harmonic oscillator equation with the minimum of the harmonic potential shifted by x_k. Thus, the wavenumber k in the y-direction determines the position of the electron in real space in the x-direction. The eigenvalues of the Schrödinger equation of an harmonic oscillator are given by $\epsilon_n^{\mathrm{HO}} = \hbar\omega_c(n + 1/2)$ with $n \in \mathbb{N}$. Hence, the energies $\epsilon_n(k)$ are actually independent of k and can be obtained via

$$\frac{\hbar^2}{2m} \left(\frac{\epsilon_n^2}{\hbar^2 v_F^2} - \frac{eB}{\hbar} \right) = \hbar\omega_c \left(n + \frac{1}{2} \right), \tag{3.23}$$

which yields the energies of LLs of Dirac systems $\epsilon_n = \pm v_F \sqrt{2\hbar e|B|n}$. These are the characteristic flat bands of LLs. The Dirac nature is reflected in the fact that the energies are proportional to $\sqrt{|B|n}$, and that there is a LL with index $n = 0$ and thus zero energy. The $n = 0$ LL, which we will refer to as lowest Landau level (LLL) from now on, is rather peculiar, since its energy is independent of the magnetic field. Such a LL does not exist for Schrödinger-like electrons.

It is convenient to redefine the LL index as $n \in \mathbb{Z}$, such that the final formula for Dirac-like LLs is given by

$$\epsilon_n = \mathrm{sgn}(Bn) v_F \sqrt{2\hbar e|B||n|}, \tag{3.24}$$

where sgn is the signum function.

Dirac electrons on a cylinder in perpendicular magnetic field

Let us now switch from the flat to the cylindrical geometry [see Fig. 3.9 b)]. In cylindrical coordinates (r, φ, z), the vector potential can be written as $\boldsymbol{A} = (0, 0, Br\sin\varphi)^T$, where we fix the radius $r = R$. In this definition, the magnetic field points into the $\hat{n}_r(\varphi = 0)$ direction, where $\hat{n}_r(\varphi)$ is the radial unit vector. We include the vector potential in the Hamiltonian for the cylindrical nanowire, given by Eq. (3.2), via minimal coupling, which yields

$$H = \hbar v_F \left[\left(\hat{k}_z + \frac{1}{\hbar} eBR\sin\varphi \right) \sigma_z + \hat{k}_\varphi \sigma_y \right]. \tag{3.25}$$

Since $\boldsymbol{A}$ is not linear in φ, the Dirac equation is not as easily solvable as before in the flat geometry. Hence, we set up a tight-binding Hamiltonian, and resort to a numerical simulation with *kwant* to compute the band structure of the infinitely long nanowire.

As an example, we choose a nanowire with a circumference of $P = 400$ nm and a magnetic field of $B = 2\,\mathrm{T}$. The absolute value of the magnetic field component perpendicular to the surface $|B_\perp(\varphi)| \equiv |B\cos\varphi|$ is maximal at the top and bottom of the nanowire shown in Fig. 3.9 b) (in our coordinates at $\varphi = 0$ and $\varphi = \pi$) and vanishes at the sides. The maximal value of the magnetic length is $l_B(\varphi = 0) \approx 18$ nm, which is much smaller than the circumference. Hence, we expect LLs to form at the top and bottom of the nanowire, and free motion at the sides.

The corresponding band structure can be seen in Fig. 3.9 a). Here, the energy of the Dirac-like LLs in flat space (with LL index $|n| < 5$ and $B = 2$ T) given by Eq. (3.24) are marked with blue dashed horizontal lines. We clearly see a flattening of the band structure at the energies of the LLs around $k_z = 0$. Most prominent is the LLL with index $n = 0$, which is perfectly flat over a large $|k_z|$ range and splits into two highly dispersing bands for $|k_z| \gtrsim 0.18\,\mathrm{nm}^{-1}$. It is two-fold degenerate in the flat region because there are two separate systems which host LLs, the top and the bottom surface of the nanowire. As for the flat geometry, a change in k_z leads to a change in the real space position of the QH state in the transversal direction. In the limit $\varphi \ll 1$, the real space position is given by $R\varphi_{k_z} = \hbar k_z/eB$ (see Eq. (3.18)), *i.e.* $k_z = 0$ corresponds to a real space position of $\varphi = 0$. For $k_z \approx 0.18\,\mathrm{nm}^{-1}$ the QH states from top and bottom surface reach the sides of the nanowire, hybridize, and disperse.

Higher LLs become less and less flat with increasing index n and finally vanish completely. Instead, the dispersion of a cylindrical nanowire without magnetic field is recovered for high energies [cf. Fig. 3.2 b)].

Let us now focus on the LL with index $n = 1$ (as an example for all LLs with $n \neq 0$). Starting from $k_z = 0$, its energy first decreases for increasing $|k_z|$. The

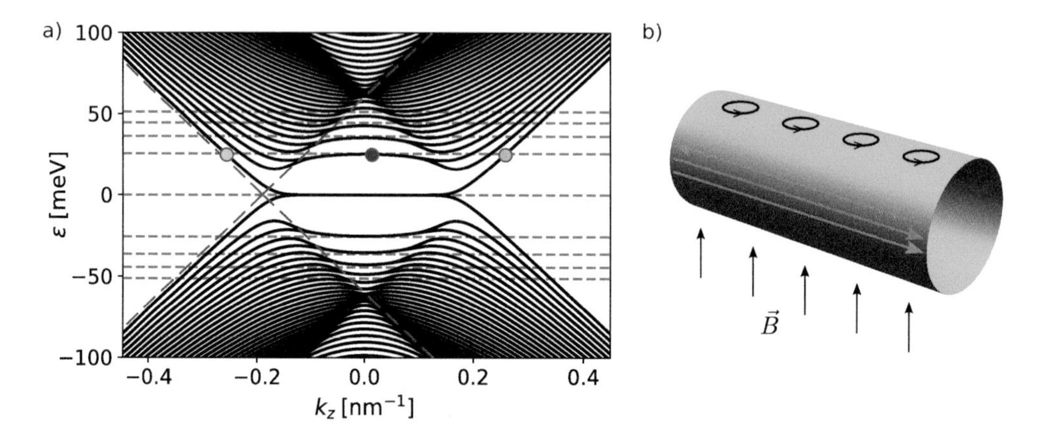

Figure 3.9: a) Band structure of a nanowire with circumference $P = 400$ nm in a perpendicular magnetic field $B = 2\,\text{T}$. The energies of Dirac LL [see Eq. (3.24)] for $|n| < 5$ are marked with dashed blue horizontal lines. The colored circles mark states whose probability distribution around the circumference is shown in Fig. 3.10. The dashed red line is a 1D Dirac cone with $v_F = 5 \cdot 10^5$ m/s added by hand. Around $k_z = 0$ LLs form, whereas for larger k_z the bands disperse and follow the 1D Dirac dispersion. b) Cylindrical nanowire in perpendicular magnetic field with QH states on top and bottom surface (black circles), and chiral side surface states (orange arrows).

reason for this is that the QH state moves to the side and therefore feels a weaker perpendicular magnetic field component $B_\perp(\varphi)$ which leads to a decrease in its energy. This effect does not occur for the LLL since its energy is independent of the magnetic field.[7] For even larger $|k_z|$, top and bottom surface states of the $n = 1$ LL hybridize at the sides (as for the LLL) and disperse strongly towards positive energy.

All of this is corroborated by Fig. 3.10, which shows the probability distribution of the angular part of the three states at the energies and wavenumbers marked by colored circles in Fig. 3.9 a). The yellow and the green circle correspond to highly dispersive modes and reside on the left and on the right side of the wire, as expected. The state marked by the purple circle, which lies within the $n = 1$ LL, resides on the bottom surface and has two maxima (the number of maxima is given by the LL index n). This state has a degenerate partner at the top surface (not shown).

The extension of QH states in real space increases with increasing LL-index $|n|$. Hence, top and bottom states start to hybridize at smaller k_z-values and the extension of the LL in k-space decreases. At energies larger than ≈ 50 meV, the

[7] Also, the Hamiltonian Eq. (3.25) respects chiral symmetry, which reads $\sigma_z H(k) \sigma_z = -H(k)$. Hence, the spectrum must be symmetric around the $\epsilon = 0$ axis.

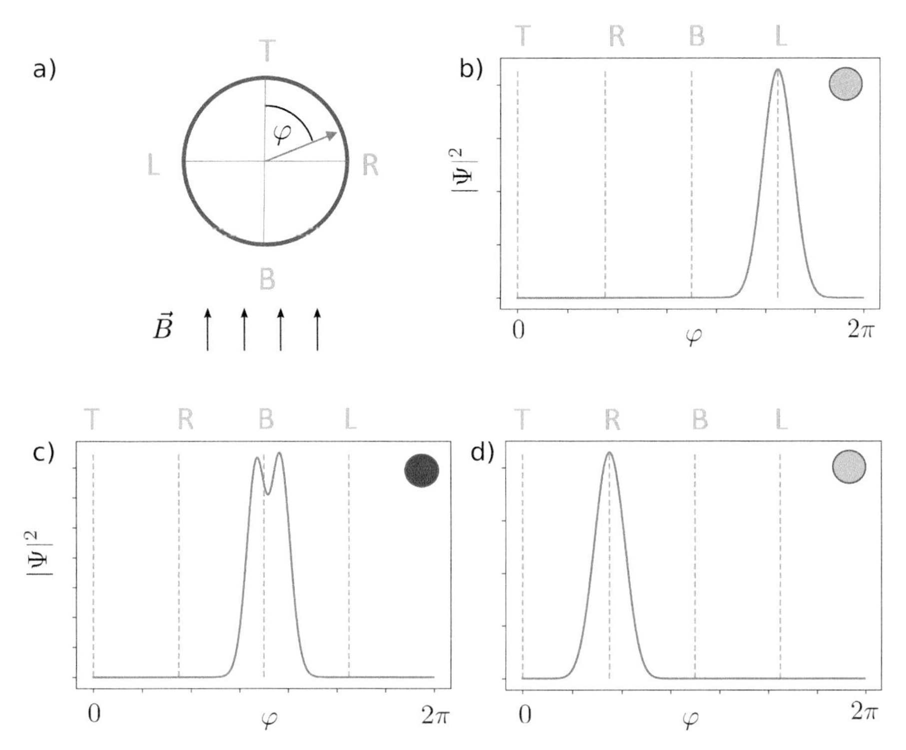

Figure 3.10: a) Cross section of the nanowire with positions top (T), right (R), bottom (B), and left (L). b-d) Probability distribution of the transversal part of the wave functions marked in Fig. 3.9 a) by colored circles. The energy of the wave functions is $\epsilon \approx 25.29$ meV, which is just below the energy of the $n = 1$ LL $\epsilon_1 \approx 25.66$ meV. The corresponding wavevectors are $k_z = -0.258\,\mathrm{nm}^{-1}$ [yellow, panel b)], $k_z = 0.0136\,\mathrm{nm}^{-1}$ [purple, panel c)], and $k_z = 0.258\,\mathrm{nm}^{-1}$ [green, panel d)].

QH states do not fit onto the top and bottom surface anymore. Therefore, the LLs vanish. Instead, for small k_z, angular momentum states wrapping around the entire circumference start to appear because the magnetic field is not strong in enough to confine the high energy states to the top or bottom surface. These states are of the same type as the states in the nanowires without magnetic field discussed in Sec. 3.1.

The final remark about the band structure concerns the slope of the dispersing modes for large k_z, which is determined by the Fermi velocity v_F. The reason for this is that $B_\perp$ vanishes at the sides of the nanowire, and we expect free Dirac electrons in 1D. Hence, part of the band structure has a shape reminiscent of a 1D Dirac cone sketched with the red dashed line in Fig. 3.9 a). The slope of this Dirac cone is given by the Fermi velocity $v_F = 5 \cdot 10^5$ m/s we used in our simulation. The highly dispersing Dirac electrons form the chiral side surface states sketched

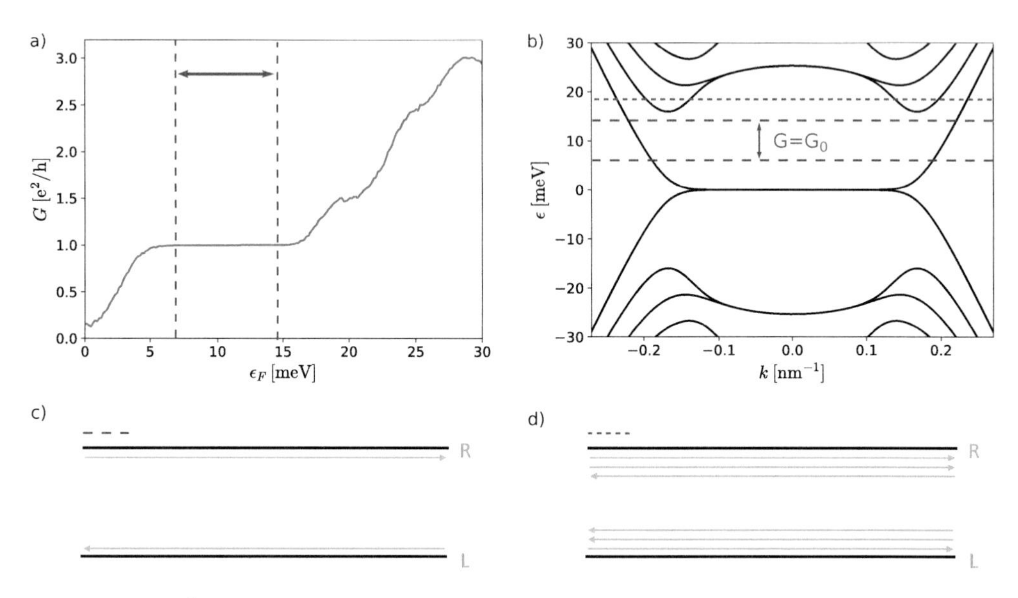

Figure 3.11: a) Disorder averaged conductance with $N_{\text{dis}} = 300$ disorder configurations through a nanowire with circumference $P = 400$ nm and length $L = 800$ nm. The perpendicular magnetic field is $B = 2$ T, the disorder strength is $K = 0.2$, and the correlation length is $\xi = 7$. A conductance plateau originating from chiral side surface states can be observed. b) Band structure of the infinitely long and clean version of the nanowire from a). c) Sketch of the top view of the nanowire with chiral side surface states in an energy window where only one mode per side is present [marked with red dashed lines in a) and b)]. d) At an energy marked with the blue dashed horizontal line in b), modes with positive and negative velocity are present on the same side of the nanowire.

with orange arrows in Fig. 3.9 b), and discussed in detail in the following.

In order to see a signature of the chiral side surface states, we simulate transport through a disordered nanowire of length $L = 800$ nm using correlated disorder and highly-doped leads. The resulting conductance as a function of the Fermi energy can be seen in Fig. 3.11 a). For easier comparison, we show the band structure of the infinite nanowire in Fig. 3.11 b) again, but this time for the relevant energy range. A prominent feature in the conductance is the flat plateau with $G = e^2/h$ within the energy range $\approx 7 - 15$ meV marked by the dashed red lines. Looking at the corresponding energy window in the band structure, we see that there is only one propagating mode per side and that those modes have opposite group velocities $v_G \equiv \partial\epsilon/\hbar\partial k$. A top view of the nanowire with one propagating mode per side is sketched in Fig. 3.11 c). Note that the situation is very similar to the QH effect. The states are topologically protected from backscattering due to the vanishing spatial overlap to their counterpropagating partner on the other side, and due to the gapped bulk.[8] This type of topological protection is fundamentally different to the one of the perfectly transmitted mode (discussed in Sec. 3.3), which comes from a vanishing spin overlap of counterpropagating modes.

For energies $\lesssim 5$ meV, the conductance drops below e^2/h. The reason for this is that scattering between the counterpropagating partners is facilitated by the disorder broadened LLL (*i.e.* by the bulk). For details about the behavior of the conductance of TI slabs in perpendicular magnetic field around zero energy (which behave qualitatively similar to cylinders), we refer to Ref. [109]. For energies $\gtrsim 15\,\text{meV}$, there are propagating and counterpropagating states on each side due to the dips in the dispersion between the flat region around $k_z = 0$ and the highly dispersive region for large $|k_z|$. A sketch of the modes for an energy marked with the dashed blue horizontal line in Fig. 3.11 b) is shown in panel d). Due to the finite spatial overlap between propagating and counterpropagating states on the same side, backscattering is possible and the conductance is susceptible to disorder.

Up to now we studied cylindrical 3DTI nanowires in perpendicular and coaxial magnetic fields. An arbitrary magnetic field is a combination of both and does not reveal any new physics. It depends on the precise strength and the direction of the magnetic field in which regime the nanowire is situated. If the magnetic length l_B computed with the maximal component of the magnetic field perpendicular to the surface is small compared to half of the circumference of the nanowire, *i.e.* $l_B \ll \pi R$, QH states confined on top and bottom surface and chiral states confined on the sides appear. Both types of states do not feel the coaxial magnetic field; hence the band structure is independent of the latter[9] and similar to the one displayed in Fig. 3.11 b). On the contrary, for $l_B > \pi R$, states wrap around the circumference and pick up the Aharonov-Bohm phase induced by the coaxial magnetic field. In this case, the band structure is similar to the one shown in Fig. 3.2 b).

[8] Here, by bulk we mean the top and bottom surface of the nanowire, *i.e.* a 2D bulk.

[9] As explained before, this is only true in an energy range where single QH states do not extend beyond the top half (or bottom half) of the wire. The crossover regime can be seen in Fig. 3.9 a) around $\epsilon \approx 60\,\text{meV}$.

4 Probing topological surface states in strained HgTe nanowires

In the last chapter, we studied the magnetotransport properties of cylindrical 3DTI nanowires using numerical simulations. We will now utilize the thereby obtained knowledge and make a transition to a real experiment based on strained HgTe nanowires. Most of the results discussed in this chapter are published in the joint experimental and theoretical work Ref. [28].

In Sec. 2.3.3, we introduced HgTe as a semimetal with an inverted band structure. According to the bulk-boundary correspondence, TSS appear at the interface between HgTe and a trivial insulator. The aim of the experiment presented in Ref. [28] is to proof that these TSS exist on the surface of HgTe nanowires (surrounded by trivial insulators) – *i.e.* to proof that strained bulk HgTe is indeed a 3DTI – by means of transport experiments.

Since HgTe is a semimetal, a bulk gap needs to be opened in order to study its surface states, which can be done for instance by applying tensile strain. Once the bulk gap is established, one can utilize the advantage that HgTe is a high mobility material with low defect doping compared to Bi-based materials, which makes residual bulk contributions negligible when studying surface transport. However, the band inversion in HgTe occurs between conduction band and the band below the valence band, hence the Dirac point is submerged in the valence band. Consequently, it is experimentally not feasible to measure the perfectly-transmitted mode (since it is masked by bulk states), which would constitute direct proof of the non-trivial topological nature of HgTe. The surface states in the experiments [28] can only be probed in an energy range where many massive subbands are occupied.

We will see, however, that it is still possible to obtain a signature of the Dirac-like surface states. A central role is thereby played by a top gate, which allows to tune the Fermi energy. We will show that a quantitative study of the conductance oscillations as a function of the gate voltage reveals that the surface states are not spin-degenerate, which is in contrast to usual Schrödinger-like states. A necessary

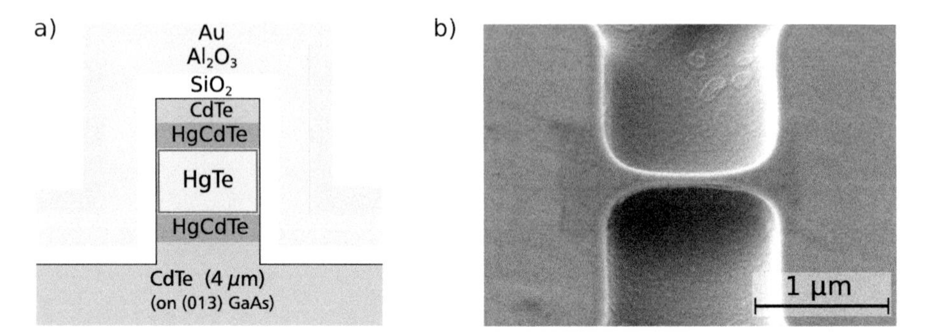

Figure 4.1: a) Sketch of a cross section of the heterostructure used in the experiments published in Ref. [28]. The embedded HgTe nanowire hosts surface states which are sketched with red color. b) Scanning electron microscope image of an etched HgTe nanowire (the wire axis runs horizontally). Adapted from [28].

ingredient for this analysis is the capacitance between gate and nanowire, which we obtain by solving the Poisson equation numerically. This analysis reveals that the gating induces a strongly inhomogeneous charge density around the wire circumference. The ramifications of the asymmetric gating on the surface state spectrum and on the transport properties of the HgTe nanowires will be an additional central topic in this chapter.

4.1 Nanowire devices

A sketch of a cross section of the heterostructures hosting a strained HgTe nanowire is shown in Fig. 4.1 a). The films on top of the (013) oriented GaAs substrates were grown by molecular beam epitaxy. The CdHgTe/HgTe/CdHgTe layers on top of the 4 μm CdTe layer, which is relaxed due to its large thickness, adapt to the lattice constant of the CdTe. The lattice mismatch between CdTe and HgTe of about 0.3% leads to a tensile strain in the 80 nm thick HgTe, which opens the desired bulk band gap, as explained in Sec. 2.3.3. With electron beam lithography and wet chemical etching nanowires were formed. A scanning electron microscope image of a representative nanowire with a length of 1.3 μm and a median width of 163 nm can be seen in Fig. 4.1 b). After etching, the nanowires were covered with Si_2O_3/Al_2O_3, and Ohmic contacts to the nanowires were formed with soldered Indium.

Surface states sketched with red color in Fig. 4.1 a) are expected to form around the HgTe nanowire due to its topologically non-trivial nature. Those surface states can be populated and depopulated by a gate which consists of gold, and which is the top layer in Fig. 4.1 a). The magnetoconductance measurements were carried out in a dilution refrigerator at temperatures between 40 mK and 300 mK. For details about the fabrication, we refer to Refs. [28] and [112].

4.2 Electrostatics of a gated nanowire – local capacitance

The electron density on the HgTe nanowire surface can be tuned by the top gate. Thereby, it is possible to position the Fermi energy inside the bulk band gap, which allows to exclusively study the surface states. Electron density n and gate voltage V_g are connected via the capacitance C between top gate and nanowire, which we define as $C \equiv en/V_g$. Note that with this definition C has units $\mathrm{F/m^2}$ as opposed to F in Sec. 2.1.

Due to the following reasons the capacitance is one of the central objects in this chapter: First, the knowledge of C allows us to compute the conductance as a function of the gate voltage $G(V_g)$ instead of as a function of the Fermi energy $G(\epsilon_F)$, which we did so far. This facilitates a direct comparison between theory and experiment. Second, the gate in the experiments presented in Ref. [28] induces a strongly non-uniform electron density around the nanowire circumference due to its shape and position. This means that the capacitance is actually position dependent, *i.e.* $C(s) = en(s)/V_g$, where s is the circumferential coordinate with $0 < s < P$, and P is the circumference of the nanowire. In this chapter, we study the influence of the non-uniformity of $n(s)$ on the band structure and the transport properties of the nanowire. Third, it is possible to obtain a signature of the Dirac surface states by a quantitative analysis of the experimentally measured conductance as a function of the gate voltage $G(V_G)$. Thereby, quantitative values for the capacitance $C(s)$ play a crucial role. Since $C(s)$ cannot be obtained from experiments [80], we resort to an electrostatic simulation discussed in the following.

In our electrostatic model, we assume that the nanowire is long such that the capacitive influence of the leads is negligible. Moreover, we neglect a varying cross section of the nanowire, and assume a perfectly rectangular cross section. Due to the first two assumptions, we only need to consider a 2D electrostatic model. Hence, we introduce Cartesian coordinates x and y in the plane of the cross section of the heterostructure [see Fig. 4.2 a)], and solve the Laplace equation in 2D for an inhomogeneous medium (different layers have different dielectric constants)

$$\nabla[\epsilon_r(x,y)\nabla u(x,y)] = 0 \tag{4.1}$$

in order to obtain the electrostatic potential $u(x,y)$. Here, $\epsilon_r(x,y)$ is the dieletric constant of the layer material present at the position (x,y). The Laplace equation

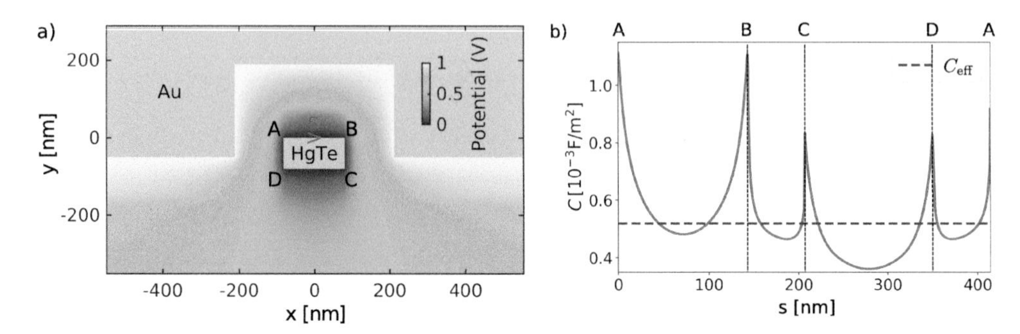

Figure 4.2: a) Cross section of the hetereostructure hosting the HgTe nanowire with electrostatic potential $u(x, y)$. The Au gate is at a potential of $u(x, y) = 1$ V and the surface of the HgTe nanowire is at $u(x, y) = 0$ V (see main text). The circumference of the wire is $P = 446$ nm and its width is $W = 163$ nm. The dielectric constants used for the different layers shown in Fig. 4.1 a) are $\epsilon_r = 10.2$ for CdTe, $\epsilon_r = 13.0$ for HgCdTe, $\epsilon_r = 3.5$ for SiO_2, and $\epsilon_r = 9.1$ for Al_2O_3. b) Capacitance as a function of the circumferential coordinate s [introduced in the main text and sketched in a)]. The effective capacitance C_{eff} marked with the dashed purple horizontal line will be introduced in Sec. 4.3.2. Adapted from [28].

is solved with the finite-element based partial-differential equation solver FENICS [113] combined with the mesh generator GMSH [114].[1] Thereby, we assume that the HgTe nanowire and the Au top gate are both perfectly metallic with vanishing electric field in the interior, which implies Dirichlet-type boundary conditions on the corresponding surfaces, *i.e.* $u(x, y) = 0$ on the HgTe nanowire and $u(x, y) = V_g$ on the boundary of the Au top gate. An example of the numerical solution of $u(x, y)$ for a device with $P = 446$ nm and for an applied gate voltage of $V_g = 1$ V can be seen in Fig. 4.2 a).

From the electrostatic potential $u(x, y)$ the surface charge density $n(s)$ on the nanowire surface can be deduced. Due to its metallic surface, the electric field is pointing along the surface normal $\hat{e}_n$ and is given by $\mathbf{E} = \hat{e}_n n(s)/\epsilon_0\epsilon_r(s)$ [116]. Here, $\epsilon_r(s)$ is the dielectric constant of the insulating layers surrounding the nanowire which are HgCdTe and SiO_2. Hence, the surface electron density can be expressed in terms of the gradient of the electrostatic potential on the nanowire surface

$$n(s) = \frac{1}{e}\epsilon_r(s)\epsilon_0[\nabla u(s)] \cdot \hat{e}_n. \tag{4.2}$$

Using Eq. (4.2), we are now able to compute the capacitance $C(s)$ between nanowire and gate as

$$C(s) = \frac{en(s)}{V_g}. \tag{4.3}$$

[1]The corresponding numerical code was kindly provided by M. H. Liu [115].

Once $C(s)$ is known, we can directly extract $n(s)$ for arbitrary gate voltage without the need for solving the Laplace equation again.

Figure 4.2 b) shows the resulting capacitance $C(s)$ for the electrostatic potential $u(x, y)$ shown in Fig. 4.2 a). The position of the edges of the nanowire are marked with letters A-D in Fig. 4.2 a) and b). The large spikes in $C(s)$ at those edges originate from the assumption that the geometry of the nanowire is perfectly rectangular, which means that it has infinitely sharp edges. This leads to an increased electric field line density at the points A-D, and hence to a capacitance value which is locally overestimated. In the experiment, the edges are smoother and thus the capacitance spikes are not expected to be that pronounced. However, we will see in Sec. 4.5 that such details are irrelevant for the magnetotransport results.

4.3 Band structure of the gated nanowire

Instead of simulating the full 3D system (*i.e.* the full heterostructure with nanowire and the gate) which is numerically very costly, we stay with the effective 2D surface model introduced in Ch. 3, and implement the effect of the gate with a position dependent onsite energy $E_{\text{gate}}(s)$ which induces the correct local electron density $n(s) = C(s)V_g/e$. Note that we extract $n(s)$ from electrostatic simulations of a nanowire with rectangular cross section but use the framework for cylindrical nanowires introduced in Ch. 3 for the transport calculations presented in subsequent sections. This is justified since the effective theory of the surface states for rectangular shaped nanowires is the same as for cylindrical nanowires [96, 97].

In order to obtain the onsite energies $E_{\text{gate}}(s)$, we need the connection between Fermi wave vector k_F and electron density n. The nanowire is a quasi 1D system with a continuous k_z but discrete k_φ, which is sketched in Fig. 3.2 a). In order to obtain the electron density n as a function of k_F, we count the states in each occupied subband labeled with index l and sum over them. The number of states N within the Fermi circle with radius k_F is then given by $N = \sum_l 2\tilde{k}_z(l)/(2\pi/L)$, where $\tilde{k}_z(l) \equiv \sqrt{k_F^2 - (2\pi l/P)^2}$ is the maximal k_z value of every subband within the Fermi circle.[2] Since the subbands we sum over reside within the Fermi circle, l is restricted to integer values which satisfy $|l| < Pk_F/2\pi$. Putting all together and adding a spin degeneracy factor g_s, the electron density for the cylindrical

[2]Here, we do not include the 1/2-angular momentum shift due to the Berry phase, *i.e.* we consider the situation where the wire is pierced by a magnetic flux $\Phi = (n + 1/2)\Phi_0$ with $n \in \mathbb{N}$. The following analysis can be readily generalized to arbitrary flux by using $\tilde{k}_z(l) \equiv \sqrt{k_F^2 - [2\pi(l + 1/2 - \Phi/\Phi_0)/P]^2}$ and by adjusting the sum over l accordingly.

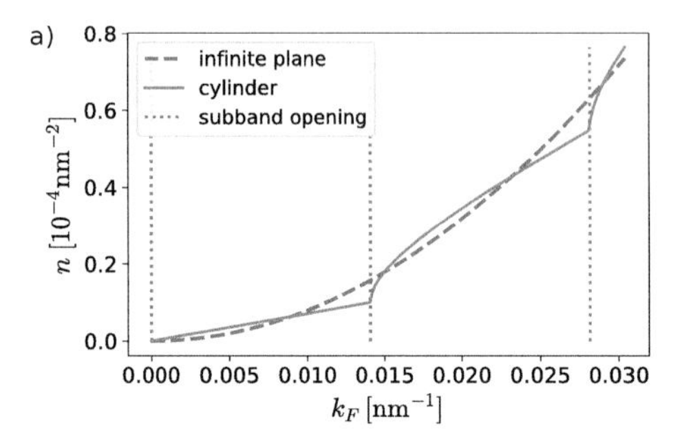

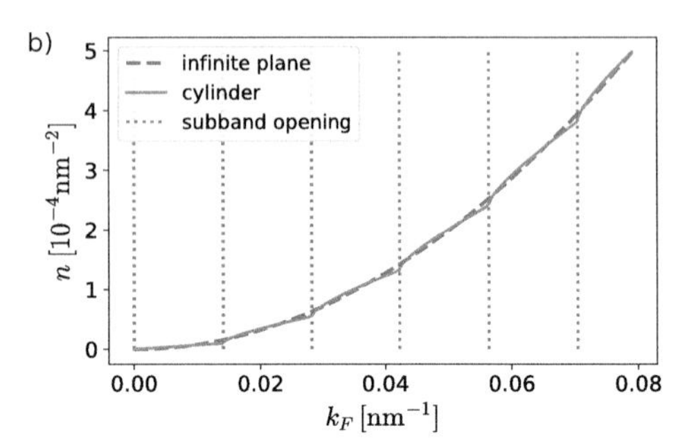

Figure 4.3: Surface electron density n as a function of the Fermi wave vector k_F for a quasi 1D system (cylinder) and a true 2D system (infinite plane). a) Within one subband the curves deviate strongly. Note that within the lowest subband (orange curve for $k_F \lesssim 0.014\text{nm}^{-1}$ we are in the true 1D limit and n increases linearly with k_F. b) On a larger scale the electron density in 2D is a good approximation for the electron density of the cylinder.

nanowire for a given Fermi wave vector k_F is given by

$$n(k_F) = \frac{g_s}{\pi P} \sum_{|l|<\frac{Pk_F}{2\pi}} \sqrt{k_F^2 - \left(\frac{2\pi l}{P}\right)^2}. \tag{4.4}$$

Note that Dirac-like surface state with a given momentum are not spin degenerate, *i.e.* $g_s = 1$, in contrast to Schrödinger electrons with $g_s = 2$. We keep g_s as a parameter in order to facilitate an easy comparison between non-degenerate surface Dirac electrons and spin-degenerate trivial surface states at a later stage in this chapter.

It is instructive to compare Eq. (4.4) to the electron density of a 2D system $n_{\text{2D}}(k_F)$ where both momenta are continuous and which reads

$$n_{\text{2D}}(k_F) = g_s \frac{k_F^2}{4\pi}. \tag{4.5}$$

In Fig. 4.3 we show both electron densities, *i.e.* we compare Eq. (4.4) with Eq. (4.5). Within one subband the shape of the curves deviates strongly, as can be seen in Fig. 4.3 a). However, on a scale large compared to the subband spacing, the electron density of the cylinder follows closely the electron density of the infinite plane. After all, in the limit of a large cylinder (or small subband spacing), one has to end up in the 2D continuum limit.

We will see later on that the shape of $n(k_F)$ between two subbands is irrelevant in order to obtain the signature of the Dirac-like surface states. Moreover, we focus on the energy window within the bulk band gap of HgTe in the experiments, which is many subbands above the Dirac point. Hence, we approximate the electron density of the quasi 1D cylinder with the electron density of a 2D continuum system in the following.

Using Eqs. (4.5) and (4.3), we are now able to induce the correct local electron density by adding the onsite energy

$$E_{\text{gate}}(s) \equiv -\hbar v_F k_F(s) = -\hbar v_F \sqrt{\frac{4\pi}{g_s} n(s)} = -\hbar v_F \sqrt{\frac{4\pi}{g_s e} V_g C(s)}, \tag{4.6}$$

where $k_F(s)$ is the local Fermi wave vector. Note that here we assume that the amount of charge on the nanowire is solely determined by the geometrical capacitance C. We merely shift the Fermi energy by the onsite energy E_{gate} such that the Fermi wave vector k_F corresponding to the correct charge density n is induced. This procedure accounts only for the Coulomb repulsion of the electrons but neglects that part of the applied gate voltage V_g is in reality used to pay for the increasing quantum single particle energies, which goes under the name of *quantum capacitance*. That this approximation is justified in our case is presented in App. A.3.

In order to obtain the band structure of the gated nanowire, we use the tight-binding model of the effective continuum Hamiltonian which describes the Dirac surface states of a cylindrical 3DTI nanowire, given by Eq. (3.7), and add the onsite term $\sum_{i,j} E_{\text{gate}}(s_j) \left|i,j\right\rangle \left\langle i,j\right| \sigma_0$. Here, j labels the transversal coordinate and $E_{\text{gate}}(s_j)$ is the discrete version of $E_{\text{gate}}(s)$ defined on the lattice. Note the connection to the cylinder is established via $R\varphi_j = s_j$. The longitudinal magnetic field is added as described in Sec. 3.3. The band structure is then computed by implementing the full tight-binding model in *kwant*.

4.3.1 Simplified capacitance model

For pedagogical reasons, we start with a simple step-shaped capacitance, depicted in Fig. 4.4, before examining the band structure of the more realistic capacitance profile shown in Fig. 4.2 b). In the simplified model, the capacitance $C(s)$ is determined by two values, one for the top surface C_{top} and one for the bottom surface C_{bot}. In the example used here, we choose $C_{\text{top}}/C_{\text{bot}} = 5$ for didactic purposes. A realistic value in the experiments in Ref. [28] is $C_{\text{top}}/C_{\text{bot}} \approx 2$.

First, we use a sketch of the resulting band structure in Fig. 4.5 a) to explain the physics behind the corresponding numerical results shown in Fig. 4.5 c). For vanishing gate voltage, the band structure is given by a simple 1D Dirac cone with quantized subbands due to the finite circumference [see left panel in Fig. 4.5 a)]. The positions of the subband minima at $k_z = 0$ are marked with blue rectangular

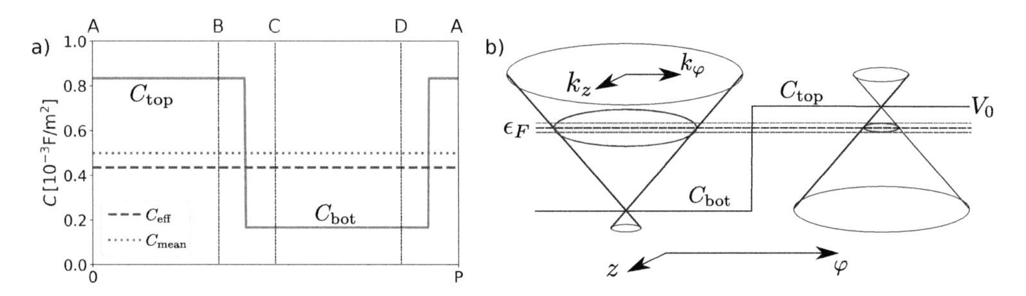

Figure 4.4: a) Simplified capacitance model with two capacitance values, one for the top surface $C_{\rm top}$ and one for the bottom surface $C_{\rm bot}$. Points A-D indicate positions along the nanowire circumference, shown in Fig. 4.2 a). Adapted from [28]. b) Gate induced potential step in the simplified capacitance model (see main text) with sketches of two Dirac cones, one on the top and one on the bottom surface. Klein tunneling is only possible if the longitudinal wavenumber k_z can be conserved, which is only the case in the darker green areas. The color code for the two colored areas is consistent with Fig. 4.5.

boxes. There is a subband minimum at zero energy (marked with a red box), since we chose a longitudinal magnetic flux of $\Phi_0/2$. For $V_g > 0$ the Dirac cone splits into two Dirac cones such that the distances to the average of the Fermi energy

$$\langle \epsilon_F \rangle = \frac{1}{P} \int_0^P \mathrm{d}s \; E_{\rm gate}(s) \tag{4.7}$$

and the two Dirac points are $\epsilon_{\rm bot} = \hbar v_F \sqrt{4\pi V_g C_{\rm bot}/e}$ and $\epsilon_{\rm top} = \hbar v_F \sqrt{4\pi V_g C_{\rm top}/e}$. Here, we assumed for convenience that the Fermi energy is at the Dirac point for $V_g = 0$. The splitting of the Dirac cone occurs since the top surface is filled faster then the bottom surface due to its larger capacitance.

Interestingly, the subband spacing at $k_z = 0$ is perfectly preserved for $V_g > 0$, which is highlighted by the equidistant rectangular boxes in the right panel of Fig. 4.5 a). This peculiar behavior can be explained with Klein tunneling [11, 117]. At $k_z = 0$ modes hit the gate induced potential step $E_{\rm gate}(s)$ associated with the capacitance profile shown in Fig. 4.4 a) perpendicularly. Hence, perfect Klein tunneling occurs, and as a consequence the modes are not affected by the inhomogeneity of the potential step. Instead, they merely feel the average of the gate induced potential $\langle \epsilon_F \rangle$.

To be more precise, consider the transversal wave vector k_φ which is not conserved any more because rotational symmetry is broken by the gate. For $k_z = 0$, it is given by the local Fermi wave vector $k_\varphi(s) = k_F(s)$ and the boundary conditions

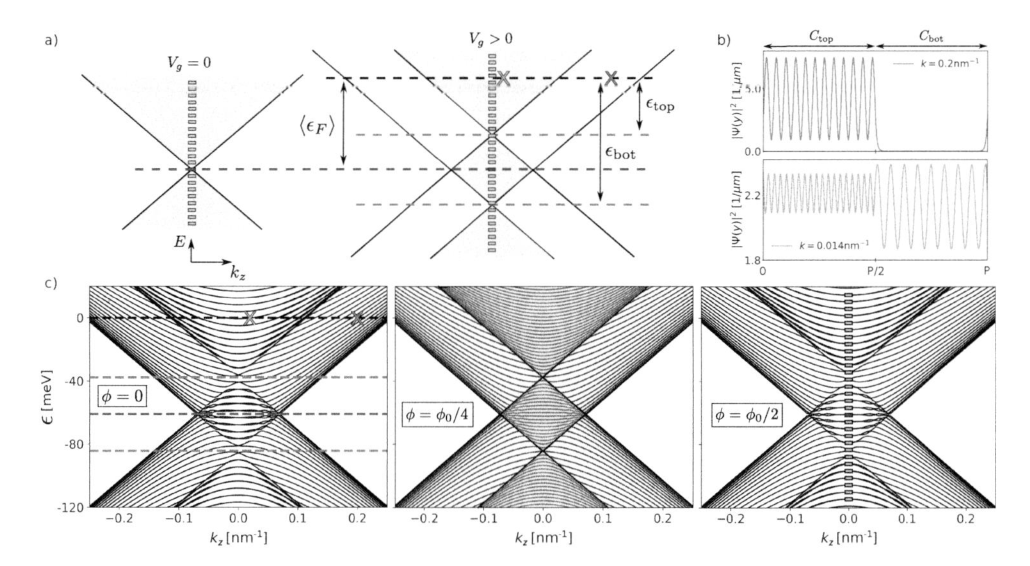

Figure 4.5: a) Sketch of the splitting of the Dirac cones for $V_g > 0$ due to the gate induced potential in the simplified capacitance model. Subband minima at $k_z = 0$ are indicated with rectangular boxes and overlapping regions of the two Dirac cones are colored with dark green. b) Probability distributions around the nanowire circumference of two representative states. The upper state (blue line) corresponds to an energy and wave number marked by a blue cross in a) and c). In the corresponding region, there are only states from the bottom Dirac cone, and thus it is confined to the bottom surface. The lower state (yellow line) corresponds to an energy and wave number marked by a yellow cross in a) and c), where an overlap between Dirac cones exists. Therefore, it is a hybridized state which lives on the top and the bottom surface. c) Simulated band structures for three magnetic fluxes $\Phi = 0$, $\Phi_0/4$, $\Phi_0/2$ (left to right panel), and a gate voltage of $V_g = 1\,\text{V}$. The Fermi velocity we use is $v_F = 5 \cdot 10^5\,\text{m/s}$. In the left panel, the distance between the black and the purple horizontal line corresponds to the average of the Fermi energy $\langle \epsilon_F \rangle$. The orange horizontal lines mark the positions of the shifted Dirac points, determined by ϵ_{top} and ϵ_{bot} (see main text). In the middle panel, areas which host flux-sensitive (insensitive) states are marked by dark (light) green color. In the right panel, subband minima at $k_z = 0$ are marked with rectangular boxes. Adapted from [28].

Eq. (3.12) imply the quantization condition

$$\exp\left[\mathrm{i}\int_0^P \mathrm{d}s\, k_\varphi(s)\right] = -\exp\left[-\mathrm{i}2\pi\Phi/\Phi_0\right], \tag{4.8}$$

where the minus sign on the right hand side originates from the curvature-induced Berry phase. It follows that the average of the transversal wave vector $\langle k_\varphi(s)\rangle$ fulfills

$$\langle k_\varphi(s)\rangle_l \equiv \frac{1}{P}\int_0^P \mathrm{d}s\, k_l(s) = \frac{2\pi}{P}\left(l + 0.5 - \Phi/\Phi_0\right), \tag{4.9}$$

i.e. it is quantized (thus we added the index l). Consequently, the subband spacing is not affected by the gate and determined by $\Delta k_l = 2\pi/P$.

For $k_z \neq 0$ there are two distinct types of states as discussed in the following. In Fig. 4.5 a), there are regions colored with dark green, where both Dirac cones overlap. A representative state of that region is shown in the lower panel of Fig. 4.5 b). It extends over the entire circumference and is thus flux sensitive. In contrast, states corresponding to regions colored in light green, where only a single Dirac cone is present (there is no overlap), are confined to the top or bottom surface as can be seen in the upper panel of Fig. 4.5 b), which shows again a representative example. Since such states do not wrap around the circumference, they are not flux sensitive. This behavior can be explained by Klein tunneling together with the conservation of the longitudinal wavenumber k_z, which is sketched in Fig. 4.4 b). It shows the gate induced potential step, whose height we denote by V_0, and the Dirac cones on top and bottom surface. Note that we consider a translationally invariant nanowire along the z-direction, *i.e.* the gate induced potential varies only in the transversal direction. The cones are filled up to the Fermi energy marked by the dashed black line. Modes from the bottom surface (large Fermi circle) can only Klein tunnel to the top surface, if there is a state with the corresponding longitudinal wavenumber k_z, which must be conserved due to the translational invariance in the z-direction. This condition is only fulfilled for states in between the two red, dashed lines. In this region states from top and bottom surface hybridize and extends around the entire circumference.

The considerations from above are clearly visible in the band structures obtained by numerical simulations, which are shown in Fig. 4.5 c) for magnetic fluxes $\Phi = 0$, $\Phi_0/4$, and $\Phi_0/2$. That is, the subband spacing is perfectly preserved, which is highlighted in the last panel of Fig. 4.5 c) by rectangular boxes. Moreover, the different flux sensitivity of the two types of states is apparent when comparing the different panels. The band structure in areas highlighted with light green do not change for different fluxes, whereas areas of the band structure highlighted with dark green vary considerably.

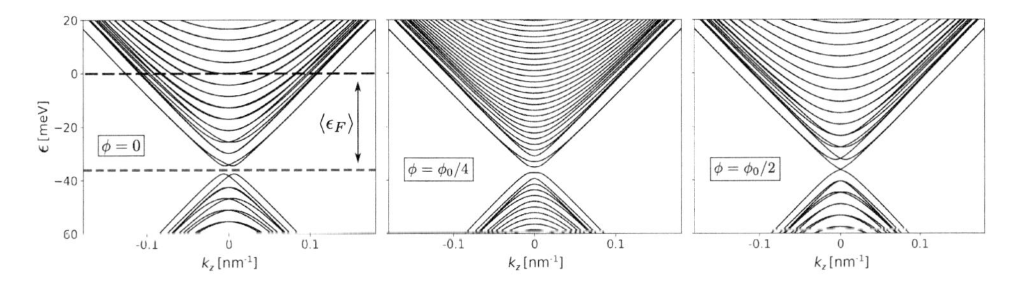

Figure 4.6: Band structures of the nanowire where the inhomogeneous charge density induced by the top gate enters via the realistic capacitance profile shown in Fig. 4.2 b) for $V_g = 0.3$ V and three magnetic fluxes. Note that the subband spacing for $k_z = 0$ for a given flux value is constant. Circumference and width of the wire are the same as in Fig. 4.2. Adapted from [28].

4.3.2 Realistic capacitance model

We are now able to analyze the band structure obtained with the more realistic capacitance profile shown in Fig. 4.2 b). The resulting band structures for three different fluxes $\Phi = 0$, $\Phi_0/4$, $\Phi_0/2$ and for a gate voltage of $V_g = 0.3$ V are shown in Fig. 4.6. The interpretation is not as simple as in the last section, since there are now four nanowire surfaces, each with a more complicated capacitance profile which is not just flat as in the simplified capacitance model. Moreover, the difference between the capacitances on the different surfaces is smaller: In the realistic scenario we are considering now, the maximal ratio between the capacitance minima on different sides is close to 1.5, whereas in the simplified capacitance model we used $C_{\mathrm{top}}/C_{\mathrm{bot}} = 5$. This means that although there are in principle four Dirac cones, one for each surface, there separation in energy is too small to recognize them. However, the main features discussed in the last section are still present. First, the perfect subband spacing at $k_z = 0$ is preserved due to Klein tunneling. Second, there are flux-sensitive and insensitive regions, although they are harder to separate from each other than in the simplified capacitance model. Note that there is a third important observation that can be made here. In the experimental relevant parameter range, the average of the Fermi energy $\langle\epsilon_F\rangle$ is always in a region where the subband minima are at $k_z = 0$. This is even fulfilled in the extreme case where $C_{\mathrm{top}}/C_{\mathrm{bot}} = 5$ [see Fig. 4.5 c)].

Let us now generalize the discussion about the average of the Fermi energy from the last section to an arbitrary capacitance profile $C(s)$. Using Eqs. (4.6) and

(4.7), we can write

$$\begin{aligned}\langle \epsilon_F(V_g) \rangle &= \hbar v_F \sqrt{4\pi V_g/e} \frac{1}{P} \int_0^P \mathrm{d}s \, \sqrt{C(s)} = \hbar v_F \sqrt{4\pi V_g/e} \left\langle \sqrt{C(s)} \right\rangle \\ &= \hbar v_F \sqrt{4\pi V_g C_{\text{eff}}/e},\end{aligned} \tag{4.10}$$

where $\langle ... \rangle$ denotes the mean value along the circumference. It is important to note that it is not the mean capacitance $C_{\text{mean}} \equiv \langle C(s) \rangle$ that enters Eq. (4.10), but rather an effective capacitance $C_{\text{eff}} \equiv \left\langle \sqrt{C(s)} \right\rangle^2$. We will see in Sec. 4.5 that it is actually the value of C_{eff} which is needed to obtain the signature of the Dirac-like surface states, and not C_{mean}. The difference between C_{eff} and C_{mean} is given by the variance of the capacitance profile $\text{Var}\left[\sqrt{C(s)}\right] = C_{\text{mean}} - C_{\text{eff}}$. For the step capacitance model with $C_{\text{top}}/C_{\text{bot}} = 5$, the values for C_{eff} and C_{mean} are plotted with horizontal lines in Fig. 4.4 a) and the ratio is $C_{\text{eff}}/C_{\text{mean}} = 0.87$. Note that in the experiments, this ratio is usually much smaller, for instance $C_{\text{eff}}/C_{\text{mean}} = 0.985$ for the realistic capacitance profile shown in Fig. 4.2 b).

4.4 Aharonov-Bohm type oscillations

In this section, we analyze the conductance as a function of a coaxial magnetic field $G(B)$ in gated HgTe nanowires in detail. Due to the Φ_0-periodicity of the spectrum of cylindrical nanowires in coaxial field [see Eq. (3.13)], we expect to see Aharonov-Bohm type oscillations in $G(B)$. Moreover, we will observe a switching between maximum and minimum of the conductance at multiples of a flux $\Phi_0/2$ depending on the gate voltage. This switching was also measured in $Bi_{1.33}Sb_{0.67}Se_3$ in Ref. [22] and in Bi_2Te_3 in Ref. [21].

In order to explain the experimental results presented in this section, it is instructive to resort to a numerical simulation of the conductance through a gated and disordered nanowire as a function of magnetic flux and gate voltage. The simulation was performed with the realistic capacitance profile obtained by an electrostatic simulation (see Sec. 4.2) and the results are presented in Fig. 4.7 b). Here, we see a $G(\Phi, V_g)$ map with a flux $0 \leq \Phi < \Phi_0$ which corresponds to one full period.[3] In the gate voltage range shown here, there are $13 - 18$ occupied two-fold degenerate subbands, *i.e.* $26 - 36$ open channels susceptible to disorder,[4] which

[3] The conductance is Φ_0-periodic since the spectrum is Φ_0-periodic.

[4] In the experiments, the number of occupied subbands can be estimated as follows. In Ref. [80], the distance between Dirac point and valence band edge in 0,3% strained HgTe is estimated to be 32 meV. With the subband spacing $\hbar v_F 2\pi/P$ this yields approximately 17 doubly degenerate subbands at the valence band edge. Note that the value 32 meV for the distance between Dirac point and valence band edge is a very rough estimate and is not conclusively discussed in the

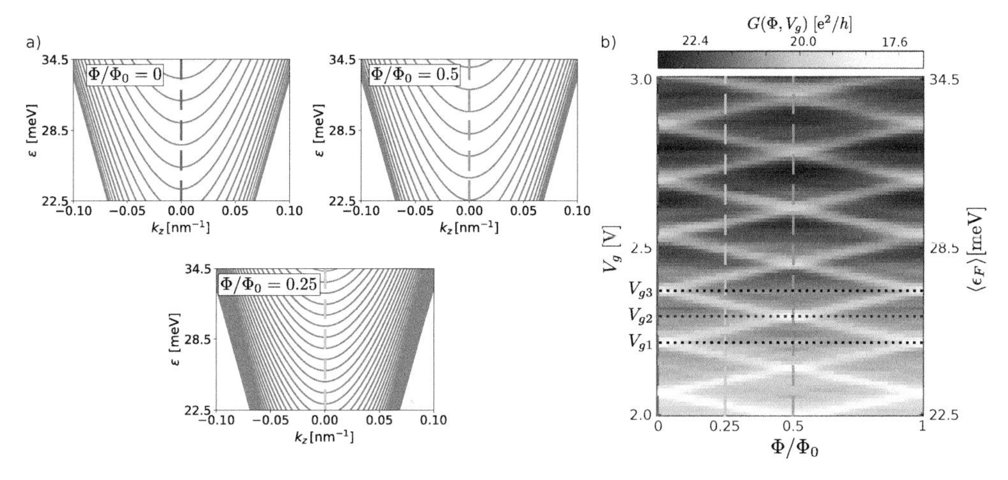

Figure 4.7: a) Band structures of a cylindrical nanowire with radius $R = 175.07$ nm without an external gate for several magnetic fluxes. b) $G(\Phi, V_g)$ map for a disordered nanowire with realistic gating. For the electrostatic model, we used a width of the wire of $W = 470$ nm and a height of $H = 80$ nm. Note that with those parameters the circumference of the wire is equal to the one used in a). Here, we used the Wilson's mass approach to avoid Fermion doubling (see Sec. 2.4.2) and Anderson disorder (see Sec. 3.2.1). The left vertical axis shows the gate voltage V_g and the right vertical axis the corresponding average of the Fermi energy $\langle\epsilon_F(V_g)\rangle$ given by Eq. (4.10). The colored vertical lines in a) and b) correspond to the same flux values. The energy of each minimum of a subband in a) can be associated with a minimum in the conductance in b) at the corresponding flux value.

explains the conductance between $\approx 17 - 23$ e$^2/h$. Overall we see the appearance of diamond like structures with lower conductance at the borders of the diamonds. The origin of those diamonds is explained in the next paragraph.

Knowledge of the value of the effective capacitance C_{eff} allows to translate the gate voltage interval $\Delta_{V_g} = [2, 3]$ V used in the simulation [left vertical axis in Fig 4.7 b)] into the energy interval $\Delta_{\langle\epsilon_F\rangle} \approx [22.5, 34.5]$ meV [right vertical axis in Fig 4.7 b)] using Eq. (4.10). Figure 4.7 a) shows the band structures of the nanowire without an external gate in the energy interval $\Delta_{\langle\epsilon_F\rangle} \approx [22.5, 34.5]$ meV for three fluxes $\Phi/\Phi_0 = 0, 0.25, 0.5$. For each flux, it is instructive to focus on the position of the subband minima at $k_z = 0$ highlighted with the vertical colored lines, and on the V_g-line cuts for the corresponding flux in Fig. 4.7 b). Whenever the average Fermi energy $\langle\epsilon_F(V_g)\rangle$ along one of the vertical line cuts in Fig. 4.7 b) reaches a local minimum in the conductance, the corresponding energy in Fig. 4.7 a) is at the bottom of a subband. As discussed in Sec. 3.3, this is due to the high

literature. However, the precise number of subbands does not play any role for the discussions to follow.

density of states (associated with a van Hove singularity) at the subband minima, which causes enhanced scattering and hence a reduced conductance. This means that the borders of the diamonds can be directly associated with minima of the corresponding subbands. Remarkably, this comparison can be done with the band structure calculated without any gate. The reason for this is Klein tunneling, which protects the subband spacings at $k_z = 0$ explained in detail in the last two sections.

Notably, the conductance minima due to van Hove singularities are the most prominent features in the conductance. Hence, we can predict the qualitative shape of the conductance as soon as C_{eff} is known. A minimum in the conductance is obtained if gate voltage and flux are chosen such that

$$\langle \epsilon_F(V_g) \rangle = \epsilon_{k_z=0,l}(\Phi) \tag{4.11}$$

is fulfilled for some angular momentum quantum number l, where $\epsilon_{k_z=0,l}(\Phi)$ is given by Eq. (3.13).

Now we are in the position to tackle the experimental results shown in Fig. 4.8. Figure 4.8 a) shows a modified conductance ΔG as a function of the magnetic flux for several gate voltages. The modified conductance ΔG is obtained by removing a small hysteresis of the superconducting magnet as well as a smoothly varying background from the experimentally measured conductance (for details see Ref. [28]). Depending on the gate voltage, there is a clear or less clear Φ_0-periodic behavior of the curves, often referred to as Aharonov-Bohm type oscillations. Moreover, a switching between maxima and minima of the conductance at fixed flux for different gate voltages can be observed. This can be explained with the flux line cuts added to Fig. 4.7 b) with dotted horizontal lines. The line cut at gate voltage V_{g1} yields a conductance minimum at zero flux and a maximum at half-integer flux. Increasing the gate voltage to V_{g2}, we observe the switching to a maximum in the conductance at zero flux and a minimum at half-integer flux. Another increase in the gate voltage to V_{g3} switches again maximum and minimum and thus brings back the situation at V_{g1}. In between those line cuts there is a transition region with no clear maximum or minimum of the conductance at multiples of $\Phi_0/2$. This might explain why at some gate voltages in Fig. 4.8 a), the maxima or minima of the conductance at multiples of $\Phi_0/2$ are less pronounced. A fast Fourier transform analysis of the magnetoconductance curves shown in Fig. 4.8 a) is presented in Fig. 4.8 b) and confirms the Φ_0-periodicity. Here, an average over fast Fourier transforms of ΔG for gate voltages 0 to 3 V was taken.

From the observations above we can draw two conclusions. First of all, the Φ_0-periodicity in the flux $\Phi = BA$ through the area A of the cross section of the nanowire as well as the phase switching between maxima and minima of the conductance are clear signs of surface transport. After all, bulk states do not

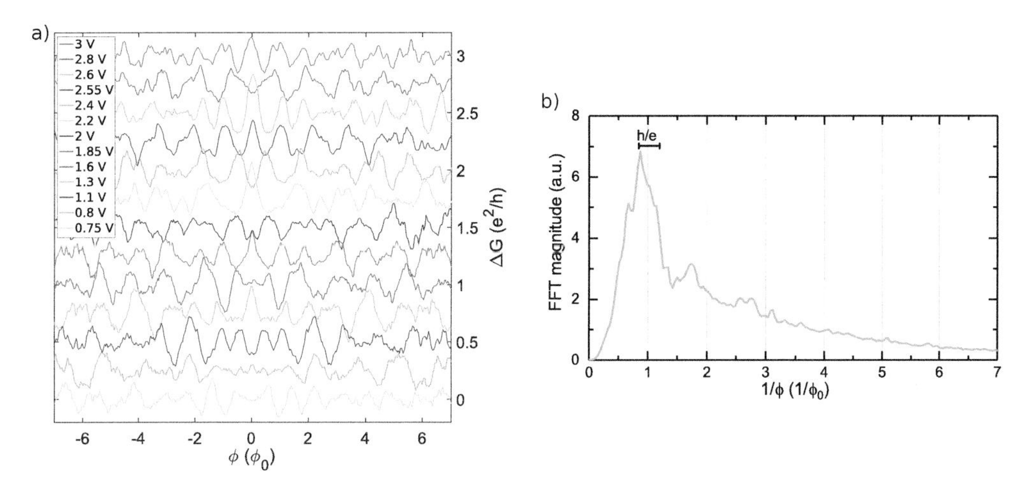

Figure 4.8: a) The modified conductance ΔG is shown for several gate voltages with an offset between each curve. The Φ_0-periodicity and the switching between minimum and maximum is evident (see main text). b) Gate voltage average of fast Fourier transforms of the magnetoconductance oscillations shown in a). A dominant peak close to $1/\Phi_0$ is observed. Adapted from [28].

care about the flux through the cross section A. Only the surface states wrap around the circumference and pick up the corresponding Aharonov-Bohm phase. We want to emphasize, however, that this behavior does not reveal the Dirac-like nature of the surface states.

Second, Φ_0-oscillations imply that the nanowires are in the non-diffusive regime, where the mean free path l_e is larger than the circumference P. In this regime, the disorder is weak enough such that the conductance is determined by the discrete nature of the subbands, which means that the subband broadening Γ is smaller than the subband spacing Δ_P, *i.e.* $\Gamma \ll \Delta_P$. In the results shown so far in this thesis, we always used disorder strengths which are small enough such that $\Gamma \ll \Delta_P$ is fulfilled. In the opposite limit, $\Gamma \gg \Delta_P$, the density of states of the system varies smoothly without any van Hove singularities, and the conductance cannot be determined by simple subband counting. The flux-dependence of the conductance is determined by weak antilocalization and the period reduces to $\Phi_0/2$ [97, 118].[5] Since there is no clear sign of a $2/\Phi_0$ peak in Fig. 4.8 b), but there are still irregular conductance fluctuations in Fig. 4.8 a), we conclude that we are in the quasi-ballistic regime, where $P < l_e < L$.

Finally, let us mention that a temperature analysis in Ref. [28] of the conductance

[5] The $\Phi_0/2$ period can be explained by the presence of TRS at multiples of $\Phi_0/2$. At those points, the destructive interference of time-reversed loops lead to an increased conductance (*i.e.* a maximum), a mechanism referred to as weak antilocalization.

oscillations (not shown) yields phase-coherence lengths l_φ larger than 1 μm and up to 5 μm in the bulk band gap. This means that l_φ is larger than the system size, which allows for a fully phase-coherent description of the system. Hence, simulating the system with *kwant*, which assumes fully phase-coherent transport, is justified.

4.5 Subband-induced conductance modulation – extraction of the spin-degeneracy

One important property which distinguishes the surface states of a TI from trivial surface states is their spin-degeneracy g_s. Trivial Schrödinger-like surface electrons are two-fold spin-degenerate, *i.e.* $g_s = 2$, whereas non-trivial Dirac-like surface states are not spin-degenerate, *i.e.* $g_s = 1$ (see Sec. 2.3.1). In this section, we discuss the quantitative analysis of the experimental conductance data which yields the spin-degeneracy g_s of the surface states, and thus allows us to draw conclusions about their very nature.

In this paragraph, we summarize all ingredients that are necessary from the last sections in order to proceed. First, we use that the coherence length l_φ is larger than the system size L ($l_\varphi \gg L$, see Sec. 4.4), which means that the system can be treated as fully coherent. Second, we make use of the fact that the disorder broadening Γ is smaller than the subband spacing Δ_P ($\Gamma \ll \Delta_P$, see Sec. 4.4), *i.e.* that the conductance is determined by the discrete nature of the subbands. The third crucial ingredient is the undisturbed subband quantization with a robust spacing in k-space of $\Delta k = 2\pi/P$ at $k_z = 0$, despite of the gate-induced, highly non-homogeneous electron density (see Secs. 4.3.1 and 4.3.2). Fourth, we use the fact that the subband minima are always located at $k_z = 0$ (see Sec. 4.3.2). The fifth ingredient we need is a reliable value for the effective capacitance C_{eff} defined in Sec. 4.3.2, which we deduce from the capacitance profile $C(s)$ obtained by an electrostatic simulation presented in Sec. 4.2.

The main idea for obtaining the spin-degeneracy g_s of the surface states is to quantitatively analyze the conductance oscillations that appear as a function of the gate voltage, which we have already seen in Fig. 4.7 b) along vertical line cuts. Because of the direct connection between the minima in the conductance and the minima of subbands, one can measure by how much the subbands fill, *i.e.* by how much the Fermi wave vector k_F increases, when the gate voltage is changed by a large enough amount such that at least two subsequent minima in the conductance appear. After all, two subsequent minima in the conductance correspond to a change in k_F from one subband to the next. Due to the connection between k_F and the electron density n [see Eq. (4.5)], the magnitude of the increase of the

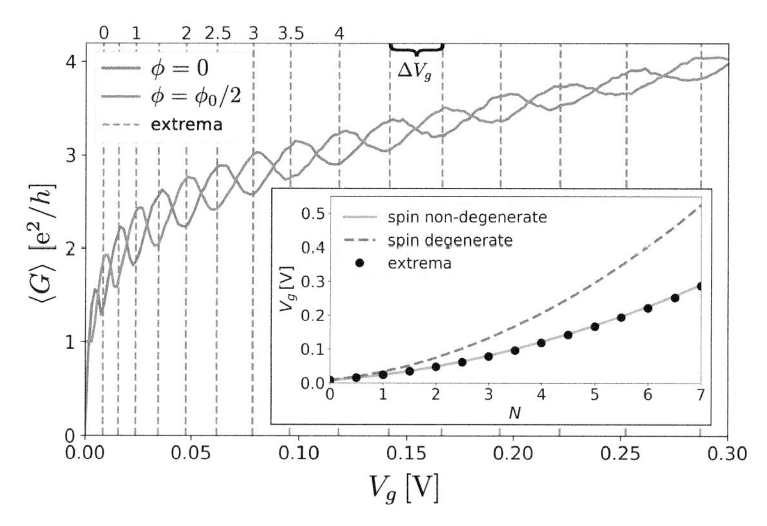

Figure 4.9: Disorder averaged conductance as a function of gate voltage for integer and half-integer magnetic flux. In the simulation 10^3 disorder configurations were used. The length of the nanowire is $L = 1.33\ \mu$m, its width is $W = 163$ nm, and its circumference is $P = 446$ nm. Correlated disorder with disorder strength $K = 0.2$ and correlation length $\xi = P/100$ was used. Note that here $V_g = 0$ corresponds to $\langle \epsilon_F \rangle = 0$, *i.e.* the Fermi energy is at the Dirac point for zero gate voltage. Minima-maxima pairs are marked with gray vertical lines and the first few are labeled with a counting index N. For $\Phi = 0/\Phi = 0.5\,\Phi_0$ minima are labeled with an integer/half-integer number. The gate-voltage of the minima-maxima pairs is plotted as a function of N in the inset of this figure. The green/purple curve shows the expected position of the minima-maxima pairs for spin non-degenerate/degenerate electrons. Adapted from [28].

Fermi wave vector Δk_F per gate voltage interval ΔV_g depends on the capacitance and on the degeneracy of the underlying states. A spin-degeneracy g_s of two means that twice as many states need to be filled compared to the non-degenerate case. Accordingly, the filling magnitude $\Delta k_F/\Delta V_g$ drops by factor of two for $g_s = 2$ compared to $g_s = 1$. By extracting $\Delta k_F/\Delta V_g$ from the conductance data $G(V_g)$, we can hence deduce the spin-degeneracy if the capacitance is known.

4.5.1 Numerical conductance data

In the following, we quantify the arguments from above by using data obtained from simulations. After that, we present an in-depth analysis of the experimental data and draw final conclusions about the nature of the surface states. Figure 4.9 shows the disorder-averaged conductance through an asymmetrically gated 3DTI nanowire as a function of gate voltage for the magnetic fluxes $\Phi = 0$ (blue curve) and $\Phi = 0.5\,\Phi_0$ (orange curve). Here, we used a realistic capacitance profile as shown in Fig. 4.2 b). Note that the anticorrelated behavior of both curves corresponds to the switching between minimum and maximum in the conductance when increasing the flux by $\Phi_0/2$, as discussed in Sec. 4.4.

As explained in Sec. 4.4, a minimum in the conductance corresponds to a position of the average of the Fermi energy $\langle \epsilon_F \rangle$ at the bottom of a subband. Due to the undisturbed subband quantization at $k_z = 0$, we can conclude that the gate voltage distance between two minima ΔV_g corresponds to an increase of the average of the Fermi wave vector $\langle k_F \rangle = \langle \epsilon_F \rangle / (\hbar v_F)$ by $\Delta k_l = 2\pi/P$ [see Fig. 3.2 a)]. Hence, the average of the Fermi wave vector at the subband minima (for a given magnetic flux) can be written as

$$\langle k_F \rangle = k_0 + N \Delta k_l, \tag{4.12}$$

where N is an integer index counting the number of filled subbands starting from a reference wave vector k_0 corresponding to the $N = 0$ minimum. Using Eq. (4.10), $\langle k_F \rangle$ can also be written as

$$\langle k_F \rangle = \sqrt{\frac{4\pi}{e g_s} C_{\text{eff}} V_g}. \tag{4.13}$$

Equating Eqs. (4.12) and (4.13), and solving for V_g leads to

$$V_g = \frac{g_s e}{4\pi C_{\text{eff}}} \left(k_0 + N \Delta k_l\right)^2, \tag{4.14}$$

which directly yields the gate voltages at which the minima in the conductance are expected.

In Fig. 4.9, the position of minima-maxima pairs of the anticorrelated curves are marked with grey vertical lines.[6] Pairs with a minimum for zero flux are labeled by an integer index, whereas pairs with a minimum for $\Phi_0/2$ flux are labeled by an half-integer index. The labeling of the minima-maxima pairs by the index N is partially shown at the top of Fig. 4.9. In the inset of Fig. 4.9, the running index N with the corresponding gate voltages V_g extracted from the minima-maxima pairs are shown (black dots). Additionally, we plot the curves defined by Eq. (4.14) for $g_s = 1$ (green solid line) and for $g_s = 2$ (purple dashed line) with a value for C_{eff} calculated from the realistic capacitance profile $C(s)$ used in the conductance simulation. The curve for Dirac-like surface states with $g_s = 1$ matches perfectly, whereas the curve for trivial surface states with $g_s = 2$ is way off. From this we can follow that the surface states in our model are spin non-degenerate. Note

[6]Note that we extracted V_g-values which are in the center between a maximum and a minimum of the anticorrelated curves instead of just taking the V_g-values of the minima. This was done to resemble the data analysis performed in the experiments, where this procedure helped to distinguish subband-induced minima from minima which originate from aperiodic conductance fluctuations. The deviations in our case are negligible.

that this is not a surprise since we used Eq. (4.6) with $g_s = 1$ to simulate the gate. However, this analysis shows that it is indeed possible to extract the spin-degeneracy by an analysis of the conductance measurements as done above despite of the highly inhomogeneous charge distribution due to asymmetric gating. The only requirement is a reliable value of the effective capacitance C_{eff}.

Before we continue with the discussion of the experimental data, we address two questions that might arise. The first question concerns the behavior of trivial Schrödinger-like surface states. Since they have a quadratic dispersion, Klein-tunneling is absent, and thus it is questionable whether the analysis from above involving Eq. (4.14) can be applied. However, an in-depth analysis presented in Sec. 4.6 shows that the conductance for Schrödinger-like electrons has a similar shape as for Dirac-like electrons in the experimentally relevant parameter range, and that the minima-maxima pairs approximately follow the spin-degenerate version of Eq.(4.14). Hence, the comparison between the curves for spin degenerate and spin non-degenerate electrons in the inset of Fig. 4.9 is justified.

The second question addresses the fact that we used the electron density as a function of the Fermi wave vector for a 2D system, given by Eq. (4.5), instead of the quasi 1D version, given by Eq. (4.4), in order to derive the gate induced onsite potential defined in Eq. (4.6). This approach is justified since we only extract gate voltages which are one subband spacing apart. In other words, we are not interested in the structure in between two subbands [see Fig. 4.3 a)], but merely probe the distances between subbands for which the 2D electron density is an excellent approximation [see Fig. 4.3 b)].

4.5.2 Experimental conductance data

We are now in the position to analyze the experimental results shown in Fig. 4.10 a). Here, the modified conductance ΔG, introduced in Sec. 4.4, is plotted as a function of the gate voltage V_g. The blue curve shows an average of $\Delta G(V_g)$ over the fluxes $\pm\Phi_0, \pm 2\Phi_0, \pm 3\Phi_0$, and the orange curve shows an average over the fluxes $\pm 0.5\Phi_0, \pm 1.5\Phi_0, \pm 2.5\Phi_0$. The averaging was done in order to decrease the influence of aperiodic conductance fluctuations. As expected from theory, the two curves show an anticorrelated behavior with minima-maxima pairs. Apart from small changes, we can apply the same procedure as we applied before for the simulated conductance data in order to extract the spin-degeneracy of the surface states. The small changes are necessary because in our simulation a vanishing gate voltage $V_g = 0$ corresponds to an average of the Fermi energy at the Dirac point with vanishing electron density $n_{\text{eff}} \equiv C_{\text{eff}} V_g / e = 0$. In the experiment, however, this is usually not the case, *i.e.* there is a non-zero surface charge density for $V_g = 0$. Moreover, we do not know the precise number of filled subbands at a

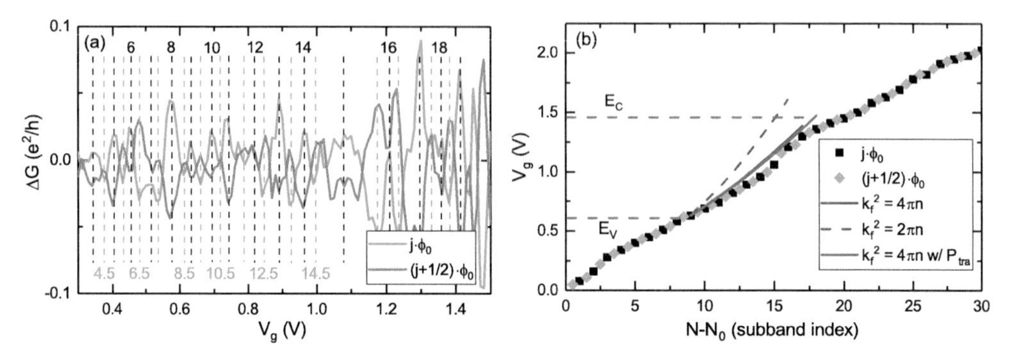

Figure 4.10: a) Modified conductance ΔG as a function of gate voltage V_g for integer (blue curve) and half-integer (orange curve) values of Φ/Φ_0. Each curve corresponds to an average over several values of the magnetic flux (see main text). The minima-maxima pairs are marked with vertical lines and labeled by a relative index $N - N_0$ (numbers on top and bottom of the vertical lines). b) Gate voltage positions of the minima-maxima pairs of a) as a function of $N - N_0$. Green diamonds correspond to green vertical lines (maximum for integer Φ/Φ_0), whereas black squares correspond to black vertical lines (maximum for half-integer Φ/Φ_0) in a). The red/blue solid lines correspond to rectangular/trapezoidal cross sections of the nanowire and show the expected behavior for Dirac-like surface states, whereas the dashed red line is associated with trivial Schrödinger-like surface states. The conductance data agrees with Dirac-like surface states. Adapted from [28].

given gate voltage V_g. Hence, we introduce a reference voltage V_0 at which N_0 subbands are filled, and a charge carrier density of n_0 is induced. This leads to slightly modified equations, which are given in the following. The connection between gate voltage and electron density reads

$$V_g - V_0 = \frac{e}{C_{\text{eff}}}(n_{\text{eff}} - n_0). \tag{4.15}$$

Using $n_{\text{eff}} = g_s \langle k_F \rangle^2/(4\pi)$ and writing $\langle k_F \rangle = k_0 + (N - N_0)\Delta k_l$ with $k_0 = \sqrt{(4\pi/g_s)(C/e)V_0}$, we obtain the equation

$$V_g - V_0 = \frac{g_s e}{4\pi C_{\text{eff}}} \left[2k_0(N - N_0)\Delta k_l + (N - N_0)^2 \Delta k_l^2\right], \tag{4.16}$$

which can be used to predict the position of the subband-induced conductance minima as a function of the relative index $N - N_0$.

Note that in the conductance simulations discussed in Sec. 4.5.1, a pure surface theory was used, *i.e.* the bulk was neglected. In the experiment, however, bulk states contribute as soon as the Fermi energy is outside of the bulk gap (for zero temperature). If this is the case, the analysis from Sec. 4.5.1 breaks down since electrons are partially filled into bulk states when the gate voltage is changed.

Thus, Eq. (4.16) is only valid within the bulk band gap. The gate voltage range corresponding to the bulk band gap in the experiment was extracted from resistance curves (not shown) [28, 81], and is marked by gray horizontal lines in Fig. 4.10 b). In the same figure, the gate voltages of the minima-maxima pairs of Fig. 4.10 a) are plotted as a function of the relative index $N - N_0$ as black squares and green diamonds. Note that V_g-values were extracted which are in the center between a minimum and a maximum of the anticorrelated curves instead of just taking the V_g-values of the minima. With this procedure, it is easier to distinguish subband-induced minima from minima which originate from aperiodic conductance fluctuations.

In order to obtain the effective capacitance of the nanowire used in the experiments, we performed electrostatic simulations (with the corresponding geometry and dielectric constants), and obtained a value of $C_{\text{eff}} = 3.987 \cdot 10^{-4}\text{F/m}^2$. With this value, we plot Eq. (4.16) for $g_s = 1$ (red solid line) and $g_s = 2$ (red dashed line) in Fig. 4.10 b). The solid blue line shows the results for $g_s = 1$ and an effective capacitance obtained for a trapezoidal cross section of the nanowire, which is closer to the real shape in the experiment, instead of a rectangular cross section. We see that both models which assume spin non-degenerate (*i.e.* Dirac-like) surface states agree very well with the values extracted from the conductance data, whereas the model for spin-degenerate states is quite off. This implies that the surface states are indeed of topological nature.

4.6 Trivial surface states

The quantitative analysis we used above to extract the spin-degeneracy of the surface states relies on Klein-tunneling, which assures an undisturbed subband spacing at $k_z = 0$. Trivial surface states, however, are expected to have a quadratic dispersion which means that Klein tunneling is absent. In this section, we answer the question whether the $G(Vg)$ oscillations with trivial surface states still allow for the quantitative analysis which results in the curve for spin-degenerate electrons in Fig. 4.10 b) (red dashed line). To this end, we resort to the band structure simulations presented in the following.

For simplicity, we assume Schrödinger-like states on a cylindrical nanowire surface without SOC described by the Hamiltonian $H = (\hat{p}_z^2 + \hat{p}_\varphi^2)/(2m^*)\sigma_0$, where m^* is an effective mass. The tight-binding version of a 1D Schrödinger Hamiltonian can be found in App. A.1. The gate enters the tight-binding Hamiltonian as an onsite energy [see Eq. (4.6)] that induces the correct electron density around the wire circumference, and reads

$$E_{\text{gate}}(s) \equiv -\frac{\hbar^2}{2m^*}k_F^2(s) = -\frac{\pi\hbar^2 C(s)}{m^* e}V_g. \tag{4.17}$$

Here, we used Eq. (4.5) with $g_s = 2$ to express the local Fermi wave vector $k_F(s)$ in terms of gate voltage V_g and capacitance $C(s)$. Due to its simplicity, we use the step-capacitance model introduced in Sec. 4.3.1, where the effect of the top gate is described by one capacitance value for the top surface C_{top} and another value for the bottom surface C_{bot}.

Figure 4.11 shows the band structure of the trivial surface states on the cylinder obtained with *kwant* for several combinations of magnetic fluxes, gate voltages, and capacitance ratios $C_{\text{top}}/C_{\text{bot}}$. The first row corresponds to uniform gating (*i.e.* $C_{\text{top}}/C_{\text{bot}} = 1$), a gate voltage of $V_g = 0.2$ V, and three magnetic fluxes $\Phi = 0$, $\Phi = 0.1\,\Phi_0$, and $\Phi = 0.5\,\Phi_0$. As expected, we see the characteristic quadratic dispersion with quadratic subband spacings of Schrödinger electrons for $\Phi = 0$. The red horizontal line corresponds to the gate induced onsite energy E_{gate} (which is position independent for uniform gating), which leads to filled bands up to the Fermi energy ϵ_F marked by the blue horizontal line. All bands are four-fold degenerate; two-fold degenerate in spin and two-fold degenerate in angular momentum. For $\Phi = 0.1\,\Phi_0$, the degeneracy between clockwise and anticlockwise movers (the degeneracy in angular momentum) splits, while it is restored for $\Phi = 0.5\,\Phi_0$. Note that all states are flux sensitive.

The second row shows the band structure for non-uniform gating with a ratio $C_{\text{top}}/C_{\text{bot}} = 2$, which is a realistic value from the experimental point of view. The gate voltage $V_g = 0.2$ V is unchanged. The gate induced onsite energies for the top/bottom surfaces are marked with green/purple dashed lines. Due to the larger capacitive coupling between top gate and top surface, the onsite energy on the top surface is lower than the onsite energy on the bottom surface [with our definition that positive gate voltage lowers the onsite energy, see Eq. (4.17)]. The gate potential induces an overall shift to negative energy overlaid by a potential well whose height is given by $E_{\text{gate}}^{\text{bottom}} - E_{\text{gate}}^{\text{top}}$. Due to the formation of this potential well, low-energy states, *i.e.* states below $E_{\text{gate}}^{\text{bottom}}$, are confined to the top surface. In contrast to Dirac-like states, this confinement happens for all longitudinal wave numbers k (in particular also for $k = 0$) since there is no Klein-tunneling. This can be seen by comparing the panel $\Phi = 0$ with $\Phi = 0.1\,\Phi_0$ in Fig. 4.11 b), which reveals that all states below $E_{\text{gate}}^{\text{bottom}}$ are not flux sensitive, *i.e.* they do not wrap around the circumference. However, states at the Fermi energy are energetically above the potential well (since the gate induces charge with the same sign on top and bottom surface), and are thus not confined to one surface. Consequently, they are flux sensitive as opposed to the low-energy states. The effect of the gate is

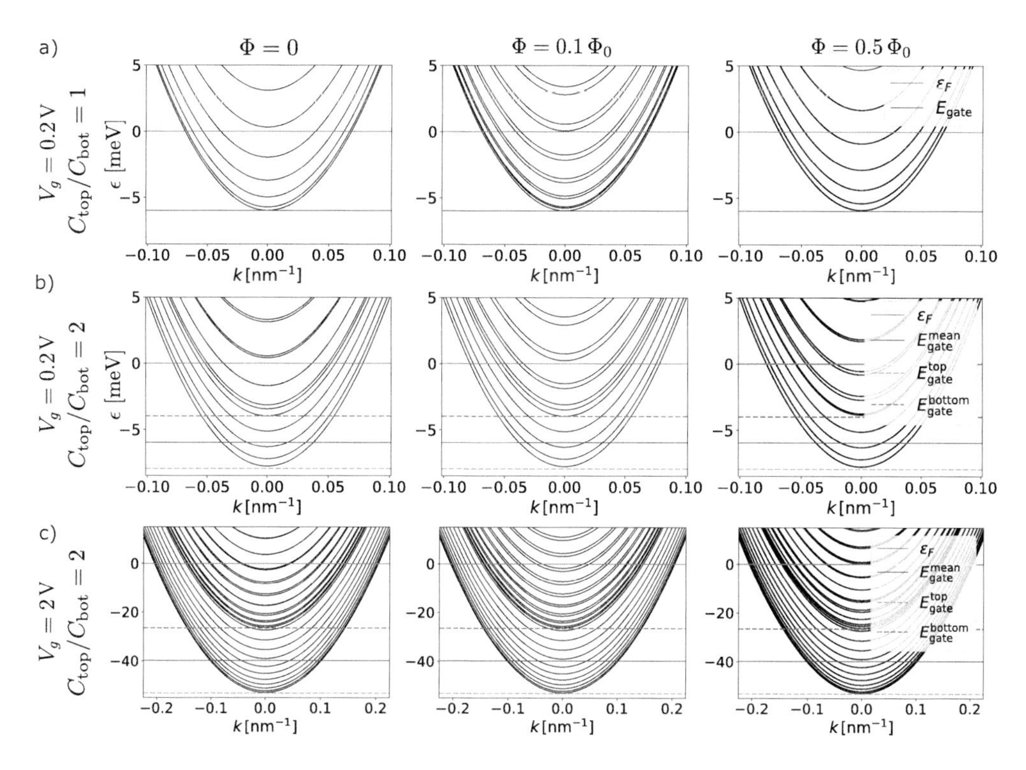

Figure 4.11: Band structure for Schrödinger-like states on a cylindrical nanowire surface with a circumference of $P = 446\,\mathrm{nm}$. Here, k is the longitudinal wave number. We assume an effective mass of $m^* = 0.03\,m_e$ (where m_e is the mass of an electron) but the qualitative form of the results does not depend on this choice. Columns correspond to magnetic fluxes $\Phi = 0$, $\Phi = 0.1\,\Phi_0$, and $\Phi = 0.5\,\Phi_0$. a) Uniform gate potential $C_{\mathrm{top}}/C_{\mathrm{bot}} = 1$ with $V_q = 0.2$ V. The bands for $\Phi = 0$ are four-fold degenerate (spin and an angular momentum degeneracy). b) Non-uniform gate potential $C_{\mathrm{top}}/C_{\mathrm{bot}} = 2$ with $V_g = 0.2$ V. The angular momentum degeneracy for $\Phi = 0$ is split by the gate induced potential. The splitting at the Fermi energy ϵ_F (blue line) is small compared to the subband spacing, and states at ϵ_F are flux sensitive while low energy states are flux insensitive. c) Non-uniform gate potential $C_{\mathrm{top}}/C_{\mathrm{bot}} = 2$ with $V_g = 2$ V. The qualitative picture is the same as in b) despite the much larger gate induced potential.

only reflected in the splitting of the angular momentum degeneracy. This splitting is, however, small compared to the subband spacing.

The observations from above also hold for much larger gate voltages, which can be seen in Fig. 4.11 c), where band structures for $V_g = 2$ V and $C_{\text{top}}/C_{\text{bot}} = 2$ are shown. There are clearly two distinct groups of states; flux insensitive states below $E_{\text{gate}}^{\text{bottom}}$, and flux sensitive states above $E_{\text{gate}}^{\text{bottom}}$ with the angular momentum degeneracy mildly split by the gate.

For the quantitative analysis which allows to extract the spin-degeneracy of the surface states, only the subband minima at the Fermi energy are of importance. Since the subband quantization at the Fermi energy is only mildly affected for experimentally relevant parameters (the lifting of the degeneracy is not within experimental precision), we expect to measure a conductance with a similar shape as for Dirac states but with minima-maxima pairs following the spin-degenerate version ($g_s = 2$) of Eq. (4.16). This was indeed confirmed by additional conductance simulations for Schrödinger-like states (not shown).

Note that, for the analysis above, we assumed that zero gate voltage corresponds to a vanishing electron density, *i.e.* we assumed that there is no intrinsic doping. Since intrinsic doping increases the subband spacing for a given gate voltage, the splitting of the degeneracy will be even less pronounced.

5 Magnetoconductance and Landau levels in topological insulator nanocones

In this chapter, we leave the terrain of nanowires with constant cross section studied so far and dive into the realm of shaped nanowires by first studying one of its simplest members, a truncated TI nanocone as sketched in Fig. 5.1 a). We will see that such a cone allows to access intriguing quantum transport phenomena, especially if magnetic fields are involved. For instance, coaxial magnetic fields $\boldsymbol{B}$ pierce the surface with a perpendicular field component $B_\perp = \boldsymbol{B} \cdot \hat{n}_\perp$ which is constant throughout the entire wire – a consequence of the linearly changing radius. We will see that such magnetic fields facilitate resonant transport through Dirac Landau levels which form around the conical surface. In general, the changing circumference along the wire axis leads to an interesting interplay between varying quantum confinement and varying magnetic flux, which results in a non-trivial mass-like potential that determines the transport properties of the cone.

Moreover, any rotationally symmetric nanowire is effectively a sequence of infinitesimal truncated cones. Hence, understanding the physics of the nanocone opens the pathway towards more exotic wire geometries. Truncated TI nanocones are also interesting from an experimental point of view: the core-shell nanowires studied in Ref. [119] have a conical shape.

5.1 Surface Hamiltonian of the nanocone

We parametrize the truncated cone by its opening angle β [see Fig. 5.1 b)] and work in the coordinates (s, φ), where φ is the azimuthal angle and s is the distance to the conical singularity (which is not part of the system) with $s_0 \leq s \leq s_1$. The corresponding local coordinate vectors $\hat{n}_\varphi$ and $\hat{n}_s$ are shown in Fig. 5.1 a). In

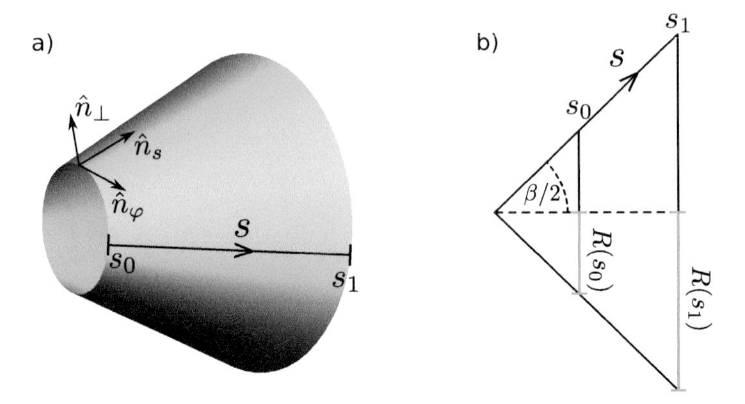

Figure 5.1: a) Sketch of a truncated conical nanowire with local coordinate vectors $\hat{n}_s$, $\hat{n}_\varphi$, and the local unit vector perpendicular to the surface $\hat{n}_\perp$. b) Side view of the cone with opening angle β and radii $R(s_0)$ and $R(s_1)$ at its ends. For the truncated cone we have $s_0 \leq s \leq s_1$, where s is the distance from the cone apex.

these coordinates, the radius is given by $R(s) = s\,\sin(\beta/2)$ and the surface Dirac Hamiltonian reads

$$H = \hbar v_F \left[\left(\hat{k}_s - \mathrm{i}\frac{1}{2s} \right) \sigma_z + \hat{k}_\varphi(s)\sigma_y \right], \tag{5.1}$$

where $\hat{k}_\varphi(s) = -\frac{\mathrm{i}}{R(s)}\partial_\varphi$. Here, the unitary transformation $\hat{U} = \exp(\mathrm{i}\sigma_z\varphi/2)$ incorporating the curvature-induced Berry phase is already applied, which means that we require antiperiodic boundary conditions for the eigenfunctions of H (see Sec. 3.1). The Dirac Hamiltonian H describes spin-1/2 particles on a space-like surface with conical geometry. Curvy-linear coordinates require to compute the spin connection [120] which leads to the additional term $-\mathrm{i}/(2s)$ compared to the Hamiltonian for the cylindrical nanowire, given by Eq. (3.2). The spin connection term ensures the Hermiticity of H with respect to the scalar product

$$\langle \Phi | \Psi \rangle \equiv \int_0^\infty \mathrm{d}s \int_0^{2\pi} \mathrm{d}\varphi\; R(s)\Phi^*(s,\varphi)\Psi(s,\varphi), \tag{5.2}$$

where the volume form $\mathrm{d}V = R(s)\mathrm{d}\varphi\mathrm{d}s$ is used. In App. A.4, the Hermiticity of H is proven for a more general geometry where the nanowire is rotationally symmetric but has an arbitrary radial profile $R(s)$. Note that H can also be derived via microscopic considerations as presented in Ref. [121].

In the following, we will show that the spin-connection term can be removed from Eq. (5.1) with a local transformation which renders the volume form trivial. Moreover, we will explain why this transformation is necessary for our numerical transport simulations with *kwant*. Tight-binding Hamiltonians are passed to *kwant* as "conventional" Hermitian matrices, *i.e.* finite-dimensional matrices which fulfill

$[(H_{\mathrm{TB}})^*]^T = H_{\mathrm{TB}}$. However, discretizing the continuum Hamiltonian Eq. (5.1) leads to a tight-binding representation which does not fulfill this condition. Let us clarify that the reason for this is the non-trivial (non-constant, *i.e.* coordinate dependent) volume form. Consider, for simplicity, a 1D system with real space coordinate q and volume form $\mathrm{d}V = f(q)dq$. On the lattice, we use the shorthand notation $f_i \equiv f(q_i)$ as well as $\Psi_i \equiv \Psi(q_i)$, where i labels the lattice sites. The condition for the Hermiticity of H, $\langle\Phi|H\Psi\rangle = \langle H\Phi|\Psi\rangle$, is then given by

$$\sum_{ij} f_i \Phi_i^* (H_{\mathrm{TB}})_{ij} \Psi_j = \sum_{ij} f_j \left[(H_{\mathrm{TB}})_{ji} \Phi_i\right]^* \Psi_j \tag{5.3}$$

on the lattice. From Eq. (5.3) it is, in general, only possible to deduce that $[(H_{\mathrm{TB}})^*]^T = H_{\mathrm{TB}}$ if $f(q)$ is constant. Hence, we use the local transformation

$$H \to \tilde{H} = \sqrt{f(q)}H\frac{1}{\sqrt{f(q)}}, \tag{5.4}$$

$$\Psi \to \tilde{\Psi} = \sqrt{f(q)}\Psi \tag{5.5}$$

which makes the volume form trivial. This can be easily seen by considering the scalar product

$$\langle\Phi|H\Psi\rangle = \int_{-\infty}^{\infty} \mathrm{d}q\, f\Phi^* H\Psi \tag{5.6}$$

$$= \int_{-\infty}^{\infty} \mathrm{d}q \left(\sqrt{f}\Phi^*\right)\left(\sqrt{f}H\frac{1}{\sqrt{f}}\right)\left(\sqrt{f}\Psi\right) \tag{5.7}$$

$$= \int_{-\infty}^{\infty} \mathrm{d}q\, \tilde{\Phi}^* \tilde{H} \tilde{\Psi}, \tag{5.8}$$

where f is fully absorbed in the wave functions and the Hamiltonian in Eq. (5.8), *i.e.* the volume form is trivial. Thus, using the transformation Eq. (5.4) yields a tight-binding Hamiltonian $\tilde{H}$ which fulfills $[(\tilde{H}_{\mathrm{TB}})^*]^T = \tilde{H}_{\mathrm{TB}}$.

Applying the transformation Eq. (5.4) to the Hamiltonian of the conical nanowire yields

$$\tilde{H} = \sqrt{R(s)}H\frac{1}{\sqrt{R(s)}} = \hbar v_F \left[\hat{k}_s \sigma_z + \hat{k}_\varphi(s)\sigma_y\right]. \tag{5.9}$$

Notably, $\tilde{H}$ has a similar form as the Hamiltonian for the cylindrical nanowire given by Eq. (3.2), except for the position dependence of the angular wavenumber $\hat{k}_\varphi$. It is this position dependence of $\hat{k}_\varphi$ where the conical geometry enters. In order to ensure the Hermiticity of the tight-binding Hamiltonian used for our simulations, we use $\tilde{H}$ instead of H in *kwant*. Moreover, due to its convenient mathematical form, we will also use $\tilde{H}$ for the analytical calculations presented in the following.

5.2 Nanocone in magnetic fields

5.2.1 Nanocone in coaxial magnetic fields – Landau levels

We add the coaxial magnetic field to the Hamiltonian defined in Eq. (5.9) via minimal coupling, which yields

$$\tilde{H} = \hbar v_F \left\{ \hat{k}_s \sigma_z + \left[\hat{k}_\varphi(s) + e A_\varphi(s)/\hbar \right] \sigma_y \right\}. \qquad (5.10)$$

Here, we used the vector potential in the symmetric gauge $\boldsymbol{A} = -BR(s)/2\,\boldsymbol{e}_\varphi \equiv A_\varphi(s)\,\boldsymbol{e}_\varphi$.[1] Since $\tilde{H}$ is rotationally symmetric, we can separate the longitudinal from the transversal coordinate with the ansatz $\tilde{\Psi}_{nl}(s,\varphi) = \mathrm{e}^{i(l+1/2)\varphi}\chi_{nl}(s)$, where $\chi_{nl}(s)$ is a two-component spinor, and l is the usual orbital angular momentum quantum number. As discussed in Sec. 3.1, the azimuthal part of $\tilde{\Psi}_{nl}(s,\varphi)$ has to fulfill antiperiodic boundary conditions due to the curvature-induced Berry phase, which results in the $+1/2$ term in the exponential. The quantum number n describes the longitudinal structure of the wave function and its precise meaning will become clear shortly. Applying $\tilde{H}$ to $\tilde{\Psi}_{nl}(\varphi, s)$ leads to the 1D Dirac equation

$$\left[v_F \hat{p}_s \sigma_z + V_l(s)\sigma_y \right] \chi_{nl}(s) = \epsilon_{nl}\chi_{nl}(s) \qquad (5.11)$$

with

$$V_l(s) \equiv \hbar v_F \frac{1}{R(s)} \left(l + \frac{1}{2} - \frac{\Phi(s,B)}{\Phi_0} \right). \qquad (5.12)$$

Here, $\Phi(s,B) = \pi R^2(s) B$ is the magnetic flux at the position s enclosed by the truncated cone. Note that at each position s, the values of $V_l(s)$ correspond to the energy eigenvalues of a Dirac electron on a ring with a magnetic flux line $\Phi(s,B)$ through the center.

In the following, we want to elucidate the significance of Eqs. (5.11) and (5.12). The Dirac operator in Eq. (5.11) is comprised of a kinetic term $v_F \hat{p}_s$ for the longitudinal motion and a position dependent mass-like term $V_l(s)$, which depends on the orbital angular momentum quantum number l. The latter term $V_l(s)$ describes the angular motion (it is the energy coming from angular momentum), and it acts as a *mass potential* (in the sense that it opens a gap) since it comes with the σ_y Pauli matrix. It is the same type of mass potential as the one given by Eq. (3.18) in Sec. 3.4, where 2D Dirac electrons in flat space subject a perpendicular magnetic field were discussed. We can conclude that Eq. (5.11) describes a 1D Dirac electron which effectively feels the absolute value of $V_l(s)$

[1] The minus sign in front of B stems from the convention for the magnetic field direction we use in this chapter, cf. inset of Fig. 5.3 b).

– it feels the *effective potential* $|V_l(s)|$. This is fundamentally different from an electrostatic potential, which enters the Dirac operator with the identity matrix, and which simply moves the Dirac cone up or down in energy. Whereas Dirac electrons Klein tunnel through electrostatic barriers, Klein tunneling through a mass potential is not possible. Hence, the effective potential $|V_l(s)|$ facilitates trapping of Dirac electrons (in contrast to electrostatic potentials), which we will discuss in detail mainly in Sec. 6.2. Moreover, we will see shortly that the effective potential $|V_l(s)|$ allows us to predict the main features of the transport characteristics of nanocones in coaxial magnetic field.

In order to develop a feeling for Eq. (5.12), let us show examples of the effective potentials $|V_l(s)|$ for a nanocone with opening angle $\beta = 15°$ and different magnetic field strengths. Figure 5.2 shows $|V_l(s)|$ for four magnetic field strengths and for all quantum numbers l relevant in the presented energy range. For clarity, only a few of them are colored and labeled. For zero magnetic field, shown in Fig. 5.2 a), the effective potential is solely determined by the centrifugal term $\propto 1/R(s)$ originating from quantum confinement. In our parametrization, the cone radius $R(s) = s\,\sin(\beta/2)$ increases with s, leading to a decreasing $|V_l(s)|$ for all l. The degenerate pairs l and $-l-1$ arise due to clockwise and anticlockwise movers with the same magnitude of the angular momentum. In Fig. 5.2 b) this degeneracy is split by a small magnetic field of $B = 0.02$ T and, on top of that, $|V_l(s)|$ is not guaranteed to decrease monotonically, as can be seen for $l = -1$. In Fig. 5.2 c) the magnetic field is increased to $B = 0.2$ T and strong deviations from the previous cases become apparent. The magnetic field term in the effective potential $|V_l(s)| = v_F\hbar\,|(l+0.5)/R(s) - BR(s)/\Phi_0|$ gives rise to positions s_i where $|V_l(s)|$ vanishes for certain angular momentum quantum numbers. In our example, those positions exist for $l = 1$ (red curve), $l = 2$ (green curve), and $l = 3$ (orange curve). Due to the absolute value, the effective potentials for those l form “wedge-shaped” quantum wells. For strong magnetic fields, as shown in Fig. 5.2 d) for $B = 2$ T, the entire plot is filled with such wedges. Here, only effective potentials for three representative angular momentum quantum numbers $l = 12, 20, 28$ are colored and labeled, which will be used in a later discussion. Note that the wedges form for l which fulfill $(l + 1/2)/R(s_0) > R(s_0)\pi B/\Phi_0$, otherwise the effective potential goes straight up.

The role of $|V_l(s)|$ in transport will be discussed in detail in Sec. 5.4. For now, let us turn to the possible bound states forming within the wedge-shaped quantum wells in Fig. 5.2 c) and d). It is possible to obtain those bound states by solving the effective 1D Dirac equation (5.11). Using the definition $\chi_{nl}(s) \equiv (f_{nl}(s), g_{nl}(s))^T$, Eq. (5.11) can be written as two first order differential equations

$$[v_F\hat{p}_s - \mathrm{i}V_l(s)]\, g_{nl}(s) = \epsilon_{nl} f_{nl}(s) \tag{5.13}$$
$$[v_F\hat{p}_s + \mathrm{i}V_l(s)]\, f_{nl}(s) = \epsilon_{nl} g_{nl}(s) \tag{5.14}$$

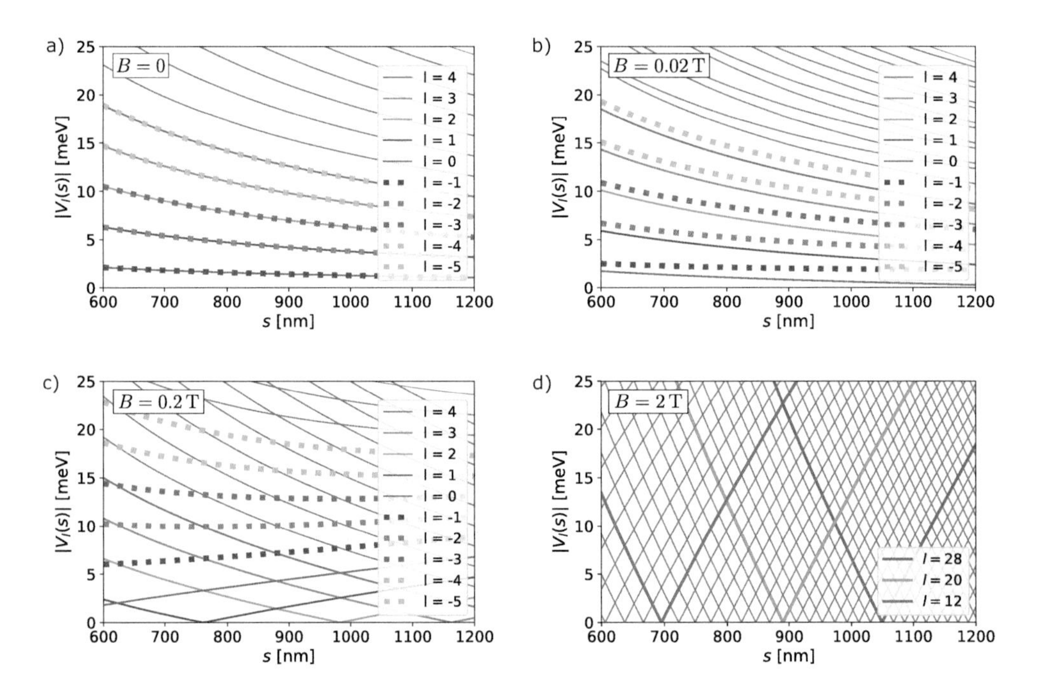

Figure 5.2: Effective potentials $|V_l(s)|$ for a nanocone with opening angle $\beta = 15°$ in the position range $s_0 = 600$ nm to $s_1 = 1200$ nm (distances measured from the conical singularity) for relevant orbital angular momentum quantum numbers l in the given energy range. $|V_l(s)|$ is colored and labeled for a few l-values, the rest is plotted with gray color. For vanishing magnetic field $B = 0$, shown in panel a), the form of $|V_l(s)|$ is solely determined by quantum confinement giving rise to the centrifugal term $\propto 1/R(s)$. For week magnetic field $B = 0.02$ T, shown in panel b), the main shape of $|V_l(s)|$ is still determined by quantum confinement but degeneracies are slightly split by the magnetic flux. For medium and especially strong magnetic fields, depicted in panels c) and d) for $B = 0.2$ T and $B = 2$ T, wedge-shaped quantum wells possibly hosting bound states form.

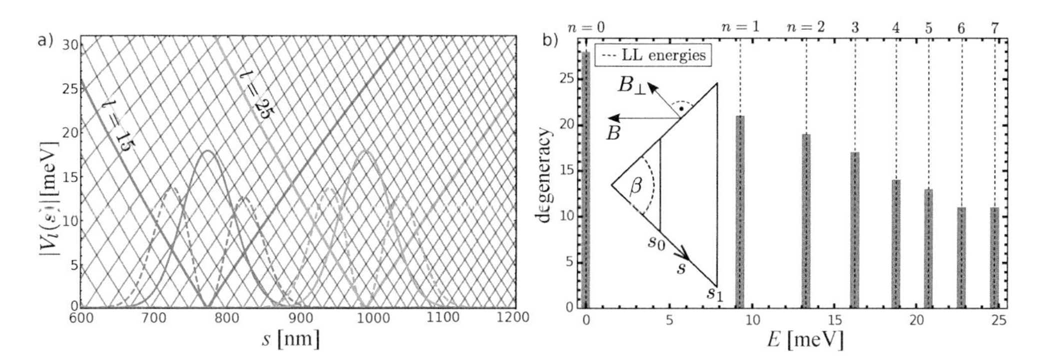

Figure 5.3: a) Effective potentials $|V_l(s)|$ for the cone parameters used in Fig. 5.2 for $B = 2$ T. Additionally, the eigenstate probability distributions [for the solutions of Eq. (5.15)] $|\chi_{n=0,l=15}(s)|^2$ (blue solid line), $|\chi_{n=1,l=15}(s)|^2$ (blue dashed line), $|\chi_{n=0,l=25}(s)|^2$ (green solid line) and $|\chi_{n=1,l=25}(s)|^2$ (green dashed line) are shown (using arbitrary units). The bound states are identified as QH states (see main text). b) Bar plot for the bound state energies ϵ_{nl}. The width of the bars is 0.4 meV. Dashed vertical lines give the analytical values of the Dirac LL energies with the perpendicular magnetic field component $B_\perp = B\sin(\beta/2)$. The inset is a sketch of the side view of the cone with coaxial magnetic field strength B and corresponding perpendicular magnetic field component $B_\perp$.

for the spinor components $f_{nl}(s)$ and $g_{nl}(s)$. Solving Eq. (5.14) for $g_{nl}(s)$ and inserting it into Eq. (5.13) yields the second order differential equation for $f_{nl}(s)$

$$\left[-\hbar^2 v_F^2 \partial_s^2 + \hbar v_F V_l'(s) + V_l^2(s)\right] f_{nl}(s) = \epsilon_{nl}^2 f_{nl}(s). \tag{5.15}$$

For a given angular momentum quantum number l, Eq. (5.15) can be solved numerically on a finite grid using Dirichlet boundary conditions far away from the wedge minimum. Then, for each effective potential wedge $|V_l(s)|$ we obtain a number of bound states which we label by the quantum number n.

Examples of probability distributions of the eigenstates $|\chi_{nl}(s)|^2$ for the cone geometry used in Fig. 5.2 with $B = 2$ T can be seen in Fig. 5.3 a) for the two wedges $l = 15$ and $l = 25$, where the solid (dashed) line corresponds to the $n = 0$ ($n = 1$) state. Here, we assumed a constant effective potential $|V_l(s)| = |V_l(s_0)|$ for $s < s_0$ and $|V_l(s)| = |V_l(s_1)|$ for $s > s_1$, which corresponds to infinite cylindrical extensions attached at s_0 and s_1. Dirichlet boundary conditions where moved away from the conical region such that convergence was achieved. Figure 5.3 b) shows a bar plot which counts the eigenenergies ϵ_{nl} in an energy window of 0.4 meV. Here, all energies ϵ_{nl} were used for which the states χ_{nl} are bound states *on* the truncated cone, *i.e.* reside between s_0 and s_1. For all ϵ_{nl} that are larger than the wedge potential at $s = s_0$ or $s = s_1$ (for instance $\epsilon_{nl} > 26$ meV for

$l = 15$) the corresponding states are not bound states on the cone but merely spurious bound states originating from the Dirichlet boundary conditions, and are thus discarded. We observe a large degeneracy in l at the Dirac LL energies $\epsilon_n = \text{sgn}(B_\perp n) v_F \sqrt{2\hbar e |B_\perp||n|}$ (up to numerical precision) derived in Sec. 3.4 and marked with vertical dashed lines. Here, $B_\perp$ is the component of the magnetic field along $\hat{n}_\perp$ in Fig. 5.1a) which is given by $B_\perp = B \sin(\beta/2)$. Hence, we can identify the bound states $\tilde{\Psi}_{nl}(s, \varphi) = \mathrm{e}^{i(l+1/2)\varphi} \chi_{nl}(s)$ of the effective potential wedges $|V_l(s)|$ as QH states. All degenerate QH states for a given n together form the n-th LL. This is consistent with the form of $|\chi_{nl}(s)|^2$, which shows one maximum for $n = 0$ and two maxima for $n = 1$.[2]

Actually, it is not a surprise that Dirac LLs form on the conical surface in the presence of a coaxial magnetic field. The perpendicular magnetic field component is constant throughout the cone, which means that the 2D Dirac electrons on the surface are subject to a homogeneous magnetic field and thus form LLs. The only condition which needs to be fulfilled is that the magnetic length is small compared to the size of the cone, *i.e.* $l_B = \sqrt{\hbar/(eB_\perp)} \ll s_1 - s_0$, such that the QH-states (in classical terms cyclotron orbits) fit onto the cone. This is equivalent to the condition that effective potential wedges form within the truncated cone. These arguments are also reflected in the degeneracies of the LLs in Fig. 5.3 b), which is given by the height of the bars. Quantum Hall states with larger n extend more in space,[3] and thus less QH states fit onto the cone. Consequently, the degeneracy decreases with increasing n.

There is a similarity between the electronic structures on the cone in coaxial magnetic field and in flat 2D space in perpendicular magnetic field, which becomes evident when comparing the QH states $\Psi_{nl}(s, \varphi)$ on the cone to the QH states in flat space $\Psi_{nk}(x, y)$ discussed in Sec. 3.4. The latter states are composed of plane waves in the y-direction with wave number k while being confined in the x-direction by a mass-potential $V_k(x)$. The position in the x direction is determined by the wavenumber k in the y direction via $x = -\hbar k/(eB)$. The case is similar for the conical nanowire: The angular wavenumber (determined by l) determines the position of the wedge $|V_l(s)|$ and hence the position in the longitudinal direction s.

We finish this section with two interesting remarks. First, looking at the cone from the front its 2D projection corresponds to a ring of finite width in a perpendicular magnetic field with strength $B_\perp$. This geometry corresponds to a Corbino disc in

[2] As we have seen in Sec. 3.4, the QH states derive from an harmonic oscillator equation. The probability distribution of an harmonic oscillator eigenstate with quantum number n has $n + 1$ maxima.

[3] Quantum Hall states derive from harmonic oscillator eigenstates, which extend more in space for larger n.

magnetic field, which is, for instance, treated in Ref. [122] using graphene. Similar to our case, resonant transmission through Dirac LLs from the outer to the inner edge was observed. Second, Eq. (5.15) is one of the standard equations appearing in supersymmetric quantum mechanics. The connection between electrons in graphene subject to a magnetic field and supersymmetric quantum mechanics is established in Ref. [123]. We expect that the connection in our case is similar.

5.2.2 Nanocone in perpendicular magnetic fields – chiral side surface states

Let us now turn to a nanocone in a magnetic field perpendicular to the nanowire axis. In our parametrization, the vector potential can be written as $\boldsymbol{A} = (0, 0, BR(s)\sin\varphi)^T$ (similar to the cylindrical nanowire in perpendicular magnetic field, see Sec. 3.4). Adding the vector potential to the Hamiltonian for the nanocone [given by Eq. (5.9)] via minimal coupling yields

$$H = \hbar v_F \left[\left(\hat{k}_s + \frac{1}{\hbar} eBR(s)\sin\varphi \right) \sigma_z + \hat{k}_\varphi(s)\sigma_y \right]. \tag{5.16}$$

Since the rotational symmetry is broken by the magnetic field, a separation ansatz as used for the cone in coaxial magnetic field cannot be used here. Moreover, translational symmetry is broken by the cone geometry. Therefore, a band structure as for the cylindrical nanowire presented in Sec. 3.4 does not exist. Hence, we resort to classical arguments to anticipate the qualitative transport behavior of the cone in perpendicular magnetic field, which we then confirm by numerical simulations in Sec. 5.4.

The analysis of the cylindrical nanowire in perpendicular magnetic field, discussed in detail in Sec. 3.4, is helpful in understanding the qualitative behavior of the surface states of the nanocone in perpendicular magnetic field. In our parametrization, the magnitude of the perpendicular magnetic field component $|B_\perp(\varphi)| = |B\cos\varphi|$ is maximal at $\varphi = 0$ and $\varphi = \pi$ (top and bottom) and it vanishes at $\varphi = \pi/2$ and $\varphi = 3\pi/2$ (sides). As long as the magnetic length $l_B = \sqrt{\hbar/[eB_\perp(\varphi = 0)]}$ is small compared to the narrow end of the cone, *i.e.* as long as $l_B \ll \pi R(s_0)$, the perpendicular magnetic field component leads to the formation of LLs on the top and bottom surfaces. On the sides, we expect free quasi-1D propagation in the form of chiral hinge states due to the vanishing perpendicular magnetic field component. In other words, the $B_\perp$-induced gap in the LL phase at the top and bottom surfaces closes at the sides where $B_\perp$ changes its sign and chiral hinge states appear.

We can conclude that, notably, the transport behavior of the cone and the cylinder in perpendicular magnetic field are expected to be the same on a qualitative level

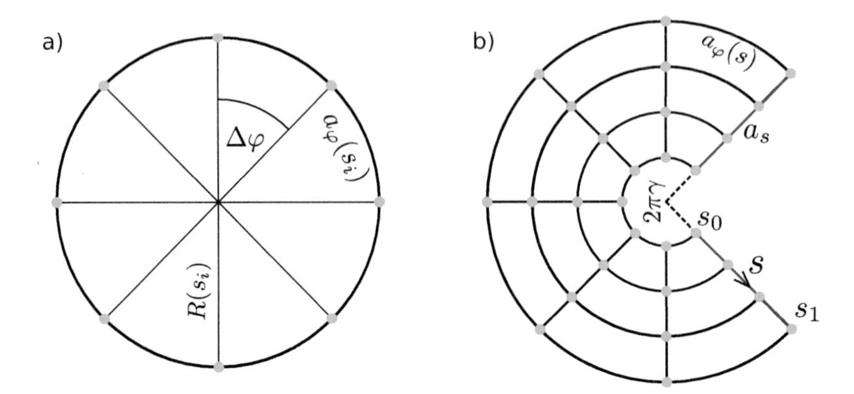

Figure 5.4: a) Numerical grid around the circumference of the cone at a position s_i. The grid spacing is given by $a_\varphi(s_i) = R(s_i)\Delta\varphi$. b) Full grid of the unfolded truncated cone. The ends (red lines) are "glued" together by setting (anti)periodic boundary conditions. The circumference $P(s)$ of the cone is given by $P(s) = 2\pi R(s) = 2\pi\gamma s$. Adapted from Ref. [124].

as long as there is enough "space" for the LLs to form on the top and bottom surface. Both are extrinsic second order TIs, whose transport properties are determined by chiral hinge states. We will see that this is in strong contrast to a cone and cylinder in coaxial magnetic field, where the precise geometry has a strong influence on the transport characteristics.

5.3 Tight-binding Hamiltonian and numerical implementation

As explained in Sec. 5.1, *kwant* takes as an input Hermitian tight-binding matrices, *i.e.* matrices which fulfill $[(H_{\mathrm{TB}})^*]^T = H_{\mathrm{TB}}$. However, discretized versions of continuum Hamiltonians with non-trivial volume form as given by Eq. (5.1) do, in general, not fulfill this requirement. Hence, we used the transformation (5.4) in Sec. 5.1 to render the volume form trivial. It turned out that this transformation also removes the term originating from the spin-connection. Hence, the final Hamiltonian we use for numerics takes the simple form given by Eq. (5.9).

In the following, we use the short-hand notation $\Psi(s_i, \varphi_j) \equiv \Psi_{i,j}$ for the two-component spinor wave function Ψ on the numerical grid defined by the grid points (i, j) (where i, j are integers). Using this notation, a discretization of the transversal wave number operator $\hat{k}_\varphi = -\mathrm{i}\partial_\varphi / R(s)$ with the standard symmetric finite difference method introduced in Sec. 2.4.1 yields

$$\hat{k}_\varphi(s)\Psi_{i,j} \to -\frac{\mathrm{i}}{R(s_i)}\frac{1}{2\Delta\varphi}\left(\Psi_{i,j+1}-\Psi_{i,j-1}\right) \tag{5.17}$$

$$\equiv -\frac{\mathrm{i}}{2a_\varphi(s_i)}\left(\Psi_{i,j+1}-\Psi_{i,j-1}\right), \tag{5.18}$$

where the angle $\Delta\varphi$ is determined by the number of grid points in the transversal direction N_φ, namely $\Delta\varphi = 2\pi/N_\varphi$. In Eq. (5.18), we introduce the s-dependent transversal grid constant $a_\varphi(s) \equiv R(s)\Delta\varphi$, which fulfills $N_\varphi a_\varphi(s) = 2\pi R(s)$. The grid along the perimeter of the cone at fixed position s_i is sketched in Fig. 5.4 a) together with the relevant parameters.

With the standard discretization of the longitudinal wave number operator $\hat{k}_s\Psi_{i,j} = -\frac{\mathrm{i}}{2a_s}\left(\Psi_{i+1,j}-\Psi_{i-1,j}\right)$ we arrive at the tight-binding Hamiltonian

$$\tilde{H}_{\mathrm{TB}} = -\frac{\mathrm{i}\hbar v_F}{2}\sum_{i,j}\left(\frac{1}{a_s}\sigma_z\left|i,j\right\rangle\left\langle i+1,j\right| + \frac{1}{a_\varphi(s_i)}\sigma_y\left|i,j\right\rangle\left\langle i,j+1\right|\right) + \mathrm{h.c.} \tag{5.19}$$

Note that the number of grid points in the transversal direction N_φ is independent of s and thus the grid spacing increases with increasing radius, which leads to a hopping integral decreasing with $\propto 1/a_\varphi(s)$. The corresponding numerical grid is sketched in Fig. 5.4 b). It shows a representation of the truncated cone in a plane, *i.e.* a truncated cone which is cut and unfolded. The cone embedded in 3D is restored by "gluing" the ends (red lines) together. The grid spacing along the s-direction is constant, whereas the transversal grid spacing increases with increasing s as $a_s(\varphi) = 2\pi R(s)/N_\varphi = 2\pi\gamma s/N_\varphi$, where $\gamma \equiv \sin(\beta/2)$. Note that $2\pi\gamma$ corresponds to the central angle in Fig. 5.4.

Leads, disorder, and local rescaling of wave functions

In order to compute transport we attach semi-infinite leads to the truncated cone, which is shown in Fig. 5.5. The leads are cylindrical extensions of the cone with radii $R(s_0)$ and $R(s_l)$. As described in Sec. 3.2.2, we add a large negative onsite energy to simulate highly-doped contacts. Gaussian correlated disorder is created with the FFM introduced in Sec. 3.2.1. We create a disorder landscape with the desired correlation length in a 3D box in which the truncated cone is embedded. The values for the disorder potential $V_{\mathrm{dis}}(\boldsymbol{r})$ are then evaluated within the box on the surface of the cone and added as an onsite potential to the tight-binding Hamiltonian (5.19). We add magnetic fields with the usual Peierls substitution as described in Sec. 3.3.

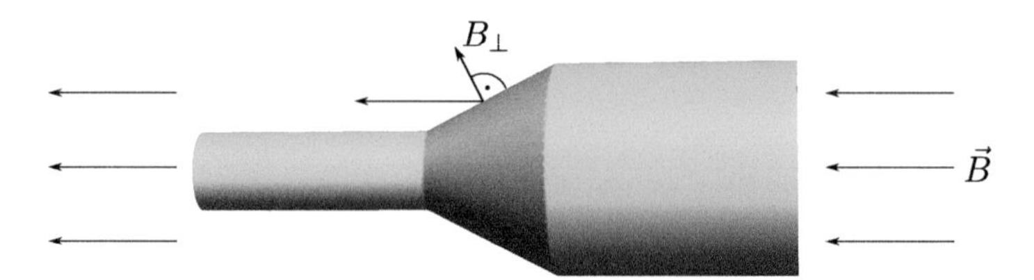

Figure 5.5: a) Sketch of a conical nanowire with cylindrical leads and coaxial magnetic field. The perpendicular magnetic field component $B_\perp$ vanishes in the leads but it is finite within the truncated cone. Adapted from Ref. [124].

Adding a coaxial magnetic field might lead to bound states (see Sec. 5.2.1), which can be computed by diagonalizing the tight-binding Hamiltonian (5.19) [apart from solving the differential equation (5.15)]. Note, however, that due to the transformation we used to render the volume form trivial [see Eq. (5.4)], the resulting wave functions $\tilde{\Psi}(s,\varphi)$ need to be transformed according to $\Psi(s,\varphi) = \tilde{\Psi}(s,\varphi)/\sqrt{R(s)}$ in order to compute physical probability densities $|\Psi(s,\varphi)|^2$. If the diagonalization is done with *kwant*, wave functions need to be rescaled according to $\Psi_{i,j} \to \Psi_{i,j}/\sqrt{a_\varphi(s_i)a_s}$ since wave functions in *kwant* are normalized such that $\sum_{i,j}\Psi^\dagger_{i,j}\Psi_{i,j} = 1$. This incorporates already the transformation $\Psi(s,\varphi) = \tilde{\Psi}(s,\varphi)/\sqrt{R(s)}$ because $a_\varphi(s) = R(s)\Delta\varphi$.

Fermion doubling

In Sec. 2.4.2 we introduced the concept of fermion doubling, which is the appearance of spurious solutions due to the discretization of an operator linear in the momentum using the central difference quotient. We extended the discussion about fermion doubling to cylindrical TI nanowires in Sec. 3.1.2. Now, the question arises to which extent and in which form fermion doubling appears in conical nanowires. The answer is not obvious – after all, we introduced fermion doubling in a system with translational symmetry, which is broken by the cone. For simplicity, we restrict ourselves to systems without magnetic fields in the following.

Let us first deal with fermion doubling in the transversal dimension, in which the cone is still translationally invariant. To this end, we use the usual ansatz $\Psi_{nl}(s,\varphi) = f(\varphi)\chi_{nl}(s)$ with $f(\varphi) = \mathrm{e}^{\mathrm{i}(l+1/2)\varphi}$ and act with the azimuthal differential operator (which is part of the Hamitlonian) in its discretized form on $f(\varphi)$, which yields

$$\partial_\varphi f(\varphi)\Big|_{\varphi=\varphi_j} = \frac{1}{2\Delta\varphi}\left[e^{i(l+1/2)(\varphi_j+\Delta\varphi)} - e^{i(l+1/2)(\varphi_j-\Delta\varphi)}\right] \tag{5.20}$$

$$= \frac{1}{\Delta\varphi}\sin\left[\left(l+\frac{1}{2}\right)\Delta\varphi\right] f(\varphi). \tag{5.21}$$

Note that the equation above is an analog of the replacement rule $k \to \sin(ka)/a$ for an infinite 1D chain with grid spacing a. Clearly, fermion doubling appears due to the discretization of the transversal momentum.

Let us now proceed with the longitudinal dimension. From Eq. (5.21), it follows that the effective 1D Dirac equation (5.11) (which we used to introduce the effective potential) for $B = 0$ is modified to

$$\left\{v_F \hat{p}_s \sigma_z + V_l^{\text{dis}}(s, B=0)\sigma_y\right\}\chi_{nl}(s) = \epsilon_{nl}\chi_{nl}(s), \tag{5.22}$$

with

$$V_l^{\text{dis}}(s, B=0) \equiv v_F \hbar \frac{1}{a_\varphi(s)} \sin\left[\frac{1}{R(s)}\left(l+\frac{1}{2}\right)a_\varphi(s)\right]. \tag{5.23}$$

A plane wave ansatz (which we used so far to derive fermion doubling) for the longitudinal part of the spinor wave function $\chi_{nl}(s)$ cannot be used since $V_l^{\text{dis}}(s, B=0)$ in Eq. (5.22) breaks translational symmetry. However, we can still argue that fermion doubling exists also for the longitudinal momentum $\hat{p}_s$. Let us assume for a moment that $V_l^{\text{dis}}(s, B=0)$ is absent. Then Eq. (5.22) simplifies to a 1D Dirac system which is translationally invariant and yields physical as well as spurious solutions. Turning $V_l^{\text{dis}}(s, B=0)$ back on affects physical and spurious solutions alike since it contains no p_s-dependence. Hence, there is no reason why the spurious solutions for the longitudinal dimension should disappear for the conical geometry. We can conclude that fermion doubling appears in both dimensions, yielding four times the number of physical states.

5.4 Magnetotransport simulations

In this section, we analyze the conductance through conical 3DTI nanowires in magnetic fields. As a representative example, we adopt the parameters used in Fig. 5.2: a truncated cone with opening angle $\beta = 15°$, $s_0 = 600\,\text{nm}$, and $s_1 = 1200\,\text{nm}$. The numerical results presented in the following were obtained using *kwant*. For details about the numerical implementation we refer to the last section.

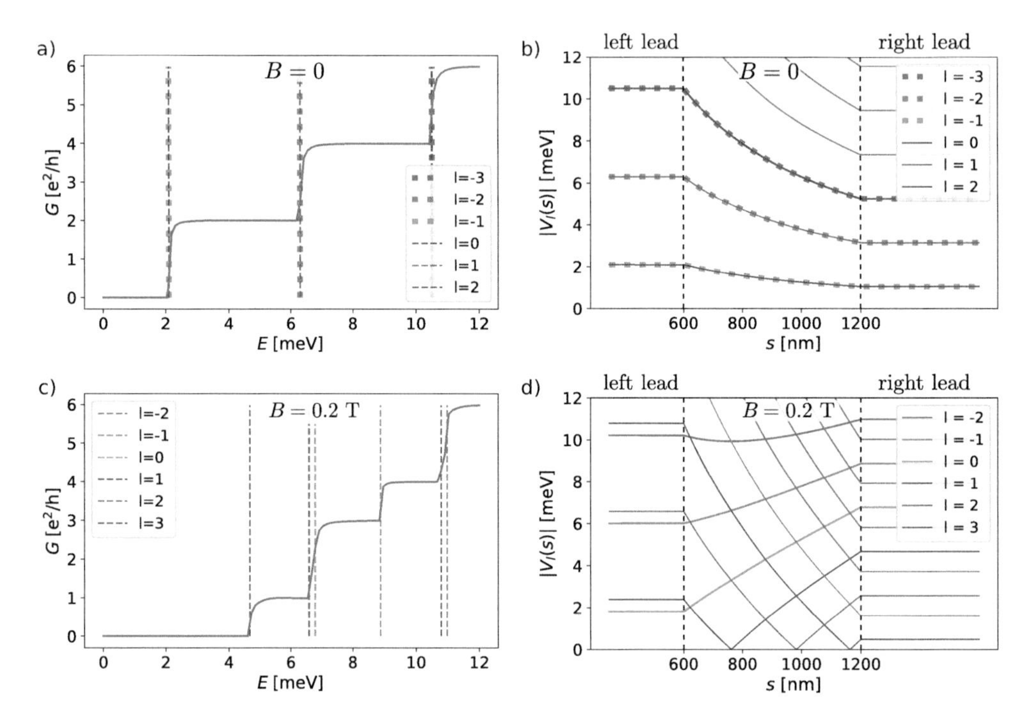

Figure 5.6: Conductance and effective potential for $B = 0$ in panels a), b) and for a coaxial magnetic field $B = 0.2$ T in panels c), d). The effective potentials are the same as in Fig. 5.2 (since the parameters of the cone are the same) but restricted to the relevant energy range and extended by the lead regions (left and right of the vertical black line). As soon as the Fermi energy is large enough such that modes with the same l-value exist in both leads, a transmission channel opens yielding a step in the conductance (see main text). The corresponding Fermi energies are marked with vertical lines in panels a) and c). In order to avoid Fermion doubling, the conductance was divided by four, which is justified since scattering (and thus inter-valley scattering) is absent without disorder.

5.4.1 Weak coaxial magnetic fields

We start our analysis by considering nanocones in magnetic fields which respect rotational symmetry. This means that we only use coaxial magnetic fields, and we do not add disorder. Due to rotational symmetry, the Dirac equation for the cone in coaxial magnetic field based on the Hamiltonian (5.10) can be written in an effective 1D form, which is given by Eq. (5.11). Here, the angular momentum acts as an effective potential $|V_l(s)|$ given by the absolute value of Eq. (5.12). In order to develop a feeling of how the effective potential affects transport, we first compute the conductance for the clean nanowire and we use leads without doping (*i.e.* without a negative onsite energy).

The conductance through the nanocone for zero magnetic field is shown in Fig. 5.6 a). We see equidistant conductance steps of the size $2e^2/h$. Their position and height can be explained with the corresponding effective potential, which is plotted in Fig. 5.6 b) for all l-values which are relevant in the depicted energy range. The central region in between the black vertical lines (600 nm $< s <$ 1200 nm) corresponds to the truncated cone. Beyond those lines we see the effective potential for the adjoining leads (cf. Fig. 5.5), which are semi-infinite cylindrical nanowires with a bands $\epsilon_{k_z,l}$ given by Eq. (3.6). In the leads, the effective potential is constant and equivalent to the subband offset $\epsilon_{k_z=0,l} = \hbar v_F \left| \frac{1}{R} \left(l + \frac{1}{2} \right) \right|$, which is the energy originating from the angular motion. Hence, the spacing of $|V_l(s)|$ is equidistant and it is larger in the left lead due to the smaller circumference. The modes which enter the scattering region (the truncated cone) via the leads are described by a well-defined orbital angular momentum quantum number l. Each of those modes feels the effective potential $|V_l(s)|$ as described in Sec. 5.2.1. As soon as the Fermi energy in the lead surpasses the effective potential, modes start to enter the cone. In the right lead, this is the case at $\epsilon_F \approx 1$ meV (for $l = -1$ and $l = 0$). However, at this Fermi energy, there are no modes available in the left lead to scatter into. Hence, the conductance stays zero until $\epsilon_F \approx 2$ meV where the corresponding modes open in the left lead. This is where the first conductance step appears in Fig. 5.6 a). For energies larger than ≈ 2 meV, a mode which enters with $l = 0$ (or $l = -1$) from the right lead gains angular momentum while traversing the cone due to the increasing effective potential, and thereby loses longitudinal momentum. Since there is no disorder, it reaches the left lead without scattering and the conductance is quantized in units of e^2/h [see Eq. (2.1)]. Due to the degeneracy of the modes for zero magnetic field the conductance steps are of the size $2e^2/h$. The energies where modes open in the left lead (and are thus open in both leads) are marked by vertical lines and colored according to their l-values in Fig. 5.6 a) [same color code as in Fig. 5.6 b)]. Note that the conductance steps at higher energies can be explained in the same way, and that the equidistant spacing between the conductance steps originates from the equidistant spacing of the subbands in the leads.

In Fig. 5.6 c) and Fig. 5.6 d) the conductance and the effective potential for a coaxial magnetic field of $B = 0.2\,\mathrm{T}$ is shown. This time, a more complicated step sequence appears with step sizes of e^2/h and $2e^2/h$. Nonetheless, this can be readily explained with the effective potential. The first step in the conductance which is of size e^2/h appears at $\epsilon_F \approx 4.75$ meV which corresponds to the opening of the $l = 1$ mode (red curve) in the right lead (in the left lead the mode opens already at $\epsilon_F \approx 2.2$ meV). At smaller energies, all modes are blocked since they are either absent in the left or in the right lead. In Fig. 5.6 c), we mark the energies where modes for a given l appear in both leads and thus lead to conductance steps. Modes with $l = 0$ (green curve) and $l = 2$ (purple curve) appear almost at the same energy around 6.75 meV. The proximity of those two mode openings leads to a step size in the conductance of $2e^2/h$.

Note that due to rotational symmetry modes cannot change their orbital angular momentum quantum number l, *i.e.* they cannot enter, for instance, with $l = 0$ on the left side and exit with $l = 3$ (brown curve) on the right side. This changes as soon as rotational symmetry is broken, for example, by disorder or asymmetric gating.

Finally, let us emphasize that for the clean cone without doping in the leads only the opening of transport channels determines the conductance. The reason for this is that the effective potential for a given l is largest in one of the leads and not within the conical region. For other geometries this need not be the case, *i.e.* maxima of the effective potential within the scattering region might appear. The effective potential then possibly acts as a tunneling barrier, which we will study in detail in Ch. 6.

5.4.2 Strong magnetic fields

When increasing the magnetic field strength to $B = 2\,\mathrm{T}$, which corresponds to a magnetic length of $l_B \approx 18.1\,\mathrm{nm}$, QH states appear on the nanocone with parameters given in the caption of Fig. 5.2. As explained in Sec. 5.2.1, these QH states correspond to bound states of the effective potentials shown in Fig. 5.2 d). In the following, we will turn to the experimentally more realistic scenario where disorder (see Sec. 3.2.1) is present and we will use highly doped leads (see Sec. 3.2.2).

Figure 5.7 a) shows the disorder-averaged conductance through the nanocone for two values of the disorder strength $K = 0.1$ (blue curve) and $K = 0.025$ (green curve). Let us first focus on the case $K = 0.025$. We see distinct conductance peaks at the position of the LL energies $\epsilon_n = \mathrm{sgn}(B_\perp n) v_F \sqrt{2\hbar e |B_\perp||n|}$ derived in Sec. 3.4 and marked by vertical orange lines. Note that the magnetic field entering here is the magnetic field component perpendicular to the surface $B_\perp = B \sin(\beta/2)$. The effective potentials $|V_l(s)|$ for the truncated cone shown in Fig. 5.7 b) offer an explanation for this peculiar conductance signature. On the first sight, each effective potential wedge is an impenetrable barrier in the given energy range, hence the conductance should be exponentially small. However, the peak height is on the order of $\simeq e^2/h$. As we have seen in Sec. 5.2.1, each of those wedges hosts bound states associated with QH states with energies ϵ_n. In Fig. 5.7 b) we sketch those bound states with horizontal orange lines for two representative wedges. All QH states with the same index n but different indices l (which are thus located in different wedges) form the LL with index n. The QH states are orthogonal –

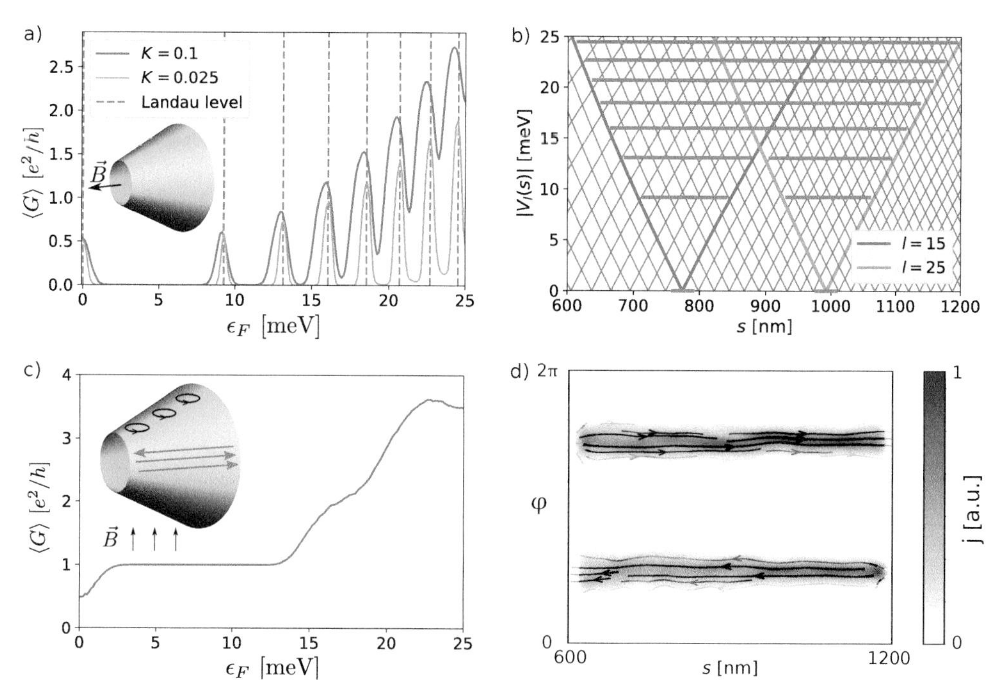

Figure 5.7: a) Disorder-averaged conductance through a nanocone in a coaxial magnetic field of $B = 2$ T (shown in the inset) for disorder strengths $K = 0.1$ and $K = 0.025$. The energies of the LLs are marked with orange vertical lines and coincide with the peaks of the conductance, apart from a small, disorder-induced shift. b) Effective potentials of the nanocone for all relevant values of the angular momentum quantum number l. The effective potentials for two representative values of l are colored and their corresponding bound states, which are QH states, are sketched with orange lines. The bound states of different wedges are degenerate and form LLs. c) Disorder-averaged conductance for a perpendicular magnetic field of $B = 2$ T and disorder strength $K = 0.1$. Transport is dominated by chiral side surface states, sketched with orange arrows in the inset of the panel. d) Current density for one chiral side surface state on each side at $\epsilon_F = 10$ meV. The arrows give the direction of the current flow. The parameters of the cone we used in this figure are the same as in Fig. 5.2. For the conductance simulations presented here, we used an average over 600 disorder configurations. The error bars, defined by Eq. (3.14), are not visible on the scale used here. In order to circumvent Fermion doubling, we used a conventional Wilson mass term as introduced in Sec. 2.4.2. Adapted from Ref. [124].

hence a transmission from one QH state to another should be forbidden. However, disorder breaks rotational symmetry, thereby coupling bound states of wedges with different l-values. Hence, the QH states are coupled, allowing electrons to pass through the cone as long as the Fermi energy in the leads is resonant with the LL energies.

A pictorial way of describing electron transport through the Dirac LLs on the cone is the following: An electron enters from one of the leads into a QH state located at the edge of the truncated cone and "jumps" from wedge to wedge until it reaches the lead on the other side. Landau levels with higher index n extend more in space, leading to a larger coupling between wedges and consequently to a larger conductance [as seen in Fig. 5.7 a)]. The broadening of the peaks can be explained by the disorder, which splits the degeneracy of the QH states in each LL due to the coupling. The coupled QH states thereby form a band – the disorder broadened LL. Stronger disorder leads to a larger broadening, which can be seen when comparing the two curves $K = 0.1$ and $K = 0.025$ in Fig. 5.7 a). Moreover, a larger disorder strength leads to stronger coupling between QH states and thus to a larger conductance, a fact which seems counterintuitive at first sight. Note that on top of the broadening, disorder shifts the LL from their original position [125, 126]. This explains the shift of the conductance peaks of the curve with disorder strength $K = 0.1$.

Let us now rotate the magnetic field by 90°, while leaving the rest of the system unchanged. As discussed in Sec. 5.2.2, we can neither resort to the effective potential (due to broken rotational symmetry) nor to a band structure (due to broken translational symmetry) to predict the transport properties of the nanocone in perpendicular magnetic field. However, for $B = 2\,\mathrm{T}$, the magnetic length is $l_B \approx 18.1$ nm which is much smaller than half of the circumference at the narrow end of the cone $2R(s_0) \approx 247$ nm. Hence, we expect LLs to form on top and bottom surface and chiral hinge states on the sides, as depicted in the inset of Fig. 5.7 c). Here, cyclotron orbits which form the LLs are sketched with black circles, while the hinge states are depicted by orange arrows. Qualitatively, the situation is the same as for the cylindrical nanowire in perpendicular magnetic field, which was discussed in Sec. 3.4. This is corroborated by the conductance through the cone, which is shown in Fig. 5.7 c). Its qualitative features are the same as those of the conductance through the cylinder depicted in Fig. 3.11 a): There is a dip at zero energy due to the coupling between chiral hinge states residing on different sides (and running in opposite directions), which is mediated by the zeroth LL [109]. The dip merges into an extended plateau with $G = e^2/h$ where transport is topologically protected. At larger energies $\epsilon_F \gtrsim 15$ meV, disorder smoothed conductance steps appear due to propagating and counterpropagating states residing on the same side. A representative current density for the chiral hinge states at $\epsilon_F = 10$ meV is shown in 5.7 d). As expected, the current runs only on the sides ($\varphi = \pi/2$ and $\varphi = 3\pi/2$) and in opposite directions.

Notably, the two settings in Fig. 5.7 a) and c) correspond to the measurement of the longitudinal and transverse conductivity, σ_{xx} and σ_{xy}, in a usual QH measurement. Using standard nomenclature, the longitudinal current density is given by $j_x = \sigma_{xx}E_x + \sigma_{xy}E_y$, where x and y denotes the longitudinal and transverse direction. Let us, for a moment, adapt this nomenclature for the nanocone. In perpendicular magnetic field, E_x vanishes as long as the Fermi energy lies within the plateau since backscattering is absent. Hence, the conductance, which is proportional to j_x, is solely determined by σ_{xy}. In longitudinal magnetic field, E_y vanishes due to the rotational symmetry of the system, which results in a conductance solely determined by σ_{xx}.

To summarize, a simple rotation of a homogeneous and sufficiently strong magnetic field from coaxial to orthogonal changes the transport properties of the nanocone drastically. In the former case, we observe non-quantized conductance peaks originating from resonant transport through Dirac LLs, while in the latter case a quantized conductance plateau originating from chiral hinge states appears. Complementary experimental transport measurements should be well within reach: Core-shell nanowires with conical shape based on HgTe have already been built [119]. Moreover, the HgTe-based nanowires introduced in Sec. 4 and studied in Ref. [28] allow to devise a spatially varying cross section profile by etching, which could be used to built nanocones. Finally, let us mention that the truly circular cross section we assumed in our model is not a necessity. In coaxial magnetic field deformations of the cone lift the degeneracies of the QH states within each LL. As long as the splitting is small compared to the broadening due to, *e.g.*, disorder or temperature, the conductance signature should not change significantly. In perpendicular magnetic field, the chiral side surface states are not affected by the details of the geometry.

6
Shaped topological insulator nanowires

Up to now, we have studied the magnetotransport properties of 3DTI nanowires with constant cross section, as well as nanowires with a linearly changing radius, *i.e.* truncated cones. Both types of nanowires have a vanishing Gaussian curvature $K = \kappa_1 \kappa_2$ on the surface, since one of the principal curvatures κ_1 and κ_2 is associated with the longitudinal direction (z- or s-coordinate), and vanishes.[1]

In this chapter, we enter for the first time the realm of curved/shaped 3DTI nanowires with nonzero Gaussian curvature, and we study the effect of the curvature on the magnetotransport properties. We will see that the concept of an effective mass potential [see Eq. (5.12)], governing the transport properties of conical nanowires in coaxial magnetic fields, can be generalized to any rotationally symmetric but otherwise arbitrarily shaped nanowire. This allows to design nanowires with almost arbitrary magnetic flux barriers, which facilitate the confinement of Dirac electrons, and thus constitute a huge playground to explore Coulomb blockade in Dirac electron systems.

While the first half of this chapter is devoted to rotationally symmetric nanowires, the second half deals with a particular wire geometry which breaks rotational symmetry, and which is closely related to the nanowires based on strained HgTe used in the experiments from Ref. [28]. We will show that magnetoconductance simulations for such wires require a special numerical grid with non-uniform grid spacings along the hopping direction. Such grids might be useful in a much broader context; hence, we explain in detail how to construct Hermitian tight-binding Hamiltonians using these grids.

Note that cylindrical nanowires with small random variations of the radius, *i.e.* with a rippled surface, have been studied in Ref. [121]. The work presented herein

[1]Note that a cone has a point-like concentration of curvature referred to as *conical singularity* at its apex, which can have a large influence on the electronic structure near the singular point [127]. Such effects are, however, outside the scope of this thesis.

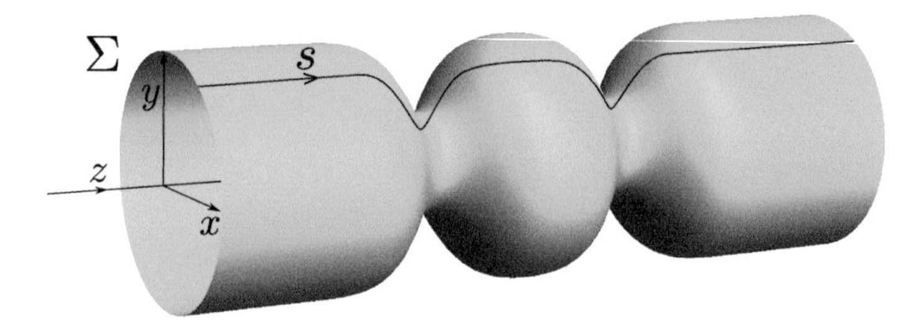

Figure 6.1: Surface of revolution Σ of a shaped rotationally symmetric nanowire with two constrictions. The arc length coordinate s and Cartesian coordinates x, y, and z are shown.

mainly concerns the robustness of the perfectly transmitted mode in the presence of the small ripples, and in combination with scalar disorder and doping.

6.1 Rotationally symmetric topological insulator nanowires

We define the surface of the nanowire Σ by a revolution of the radial profile $\tilde{R}(z)$ around the Cartesian axis z, depicted in Fig. 6.1. It is convenient to switch to the arc length coordinate

$$s(z) = \int_{z_0}^{z} \mathrm{d}z' \sqrt{1 + \left(\frac{\mathrm{d}\tilde{R}}{\mathrm{d}z'}\right)^2} \tag{6.1}$$

with the corresponding radial profile $R(s) = \tilde{R}(z(s))$. We describe a point on Σ with the coordinates (s, φ), which are connected to Cartesian coordinates via

$$\boldsymbol{r} \equiv \begin{pmatrix} x \\ y \\ z \end{pmatrix} = \begin{pmatrix} R(s)\cos\varphi \\ R(s)\sin\varphi \\ z(s) \end{pmatrix}, \tag{6.2}$$

where $z(s)$ is the inverse of Eq. (6.1). Note that the surface Σ comes with the volume form

$$\mathrm{d}V = R(s)\mathrm{d}\varphi\mathrm{d}s, \tag{6.3}$$

which is derived in App. A.4.

In these coordinates, the Dirac Hamiltonian on Σ takes the form

$$H = \hbar v_F \left[\sigma_z \left(\hat{k}_s - \frac{\mathrm{i}}{2}\frac{R'(s)}{R(s)}\right) + \sigma_y \hat{k}_\varphi(s)\right], \tag{6.4}$$

where $\hat{k}_\varphi(s) = -\frac{\mathrm{i}}{R(s)}\partial_\varphi$. The Hamiltonian H above is a generalization of the Dirac Hamiltonian on a conical surface, given by Eq. (5.1). The term originating from the spin-connection [120] is now given by $-\mathrm{i}R'(s)/[2R(s)]$, which equals $-\mathrm{i}/(2s)$ for the conical geometry. The spin connection term takes care of the fact that while the electron moves on Σ, the angle between surface normal and spin quantization axis is fixed. The motion of the spin is thus described by spin adiabatic parallel transport. Moreover, the spin connection term ensures the Hermiticity of H, which is proven in App. A.4. Note that the unitary transformation $U = \exp(i\sigma_z\varphi/2)$ is already applied, *i.e.* we require antiperiodic boundary conditions for the eigenfunctions of H (see Sec. 3.1). A derivation of H based on microscopic considerations is presented in Ref. [121].[2]

As explained in Sec. 5.1, a tight-binding version of a Hamiltonian with non-trivial volume form does in general not fulfill $[(H_{\mathrm{TB}})^*]^T = H_{\mathrm{TB}}$, which is, however, necessary for our numerical simulations. Hence, we render the volume form trivial with the local transformation $H \to \tilde{H} = \sqrt{R(s)}H/\sqrt{R(s)}$ [see Eq. (5.4)], which yields

$$\tilde{H} = \hbar v_F \left[\sigma_z \hat{k}_s + \sigma_y \hat{k}_\varphi(s)\right], \tag{6.5}$$

where we used the identity

$$\sqrt{R(s)}\left(\partial_s + \frac{R'(s)}{2R(s)}\right)\frac{1}{\sqrt{R(s)}} = \partial_s. \tag{6.6}$$

Notably, the spin-connection term is removed by the transformation, as in the case of the conical nanowire. We are left with the simple Dirac Hamiltonian given by Eq. (6.5), where the entire geometry enters via $R(s)$ in the angular wave number operator $\hat{k}_\varphi(s) = -\frac{\mathrm{i}}{R(s)}\partial_\varphi$.

Coaxial magnetic field

As usual, we add a coaxial magnetic field via minimal coupling and separate the azimuthal part of the wave function from the longitudinal part with the ansatz $\Psi_{nl}(s,\varphi) = \mathrm{e}^{i(l+1/2)\varphi}\chi_{nl}(s)$, as we have done for the nanocone in Sec. 5.2.1. Due to the same form of the Hamiltonian for the nanocone and the curved nanowire, cf. Eqs. (5.10) and (6.5), we arrive at the same effective 1D Dirac equation (5.11) with the angular momentum mass potential $V_l(s)$ defined in Eq. (5.12) when applying $\tilde{H}$ to $\Psi_{nl}(s,\varphi)$. The only difference is the radius $R(s)$, which is now an

[2] In Ref. [121], H is expressed in terms of the Cartesien coordinate z instead of the arc length coordinate s. However, using $\mathrm{d}s = \sqrt{1 + (\mathrm{d}\tilde{R}/\mathrm{d}z)^2}\mathrm{d}z$, the connection is straight forward.

arbitrary function as opposed to the linear function $R(s) = \gamma s$ for the cone. The concept of the angular momentum term acting as an effective potential

$$|V_l(s)| = v_F \hbar \frac{1}{R(s)} \left| l + \frac{1}{2} - \frac{\Phi(s,B)}{\Phi_0} \right| \tag{6.7}$$

first introduced in Sec. 3.4 in the context of 2D Dirac electrons in flat space in a homogeneous magnetic field, and then extended to nanocones in coaxial magnetic field in Sec. 5.2.1, also holds for nanowires with generic radial functions $R(s)$. Depending on its angular momentum quantum number l, each electron feels a distinct effective potential $|V_l(s)|$ which determines its behavior within the curved nanowire and hence the transport properties of the latter.

6.1.1 Magnetotransport simulations

The Hamiltonian $\tilde{H}$, given by Eq. (6.5), can be discretized in the same way as the Hamiltonian for the nanocone, which was presented in Sec. 5.3. The resulting tight-binding representation is given by Eq. (5.19) with $a_\varphi(s) = 2\pi R(s)/N_\varphi$, where N_φ is the number of grid points in the azimuthal direction. Note, however, that as opposed to the cone, an unfolded version of a curved nanowire with arbitrary radial profile $R(s)$ does not exist. Concerning Fermion doubling, the same line of arguments as for the nanocone holds (see Sec. 5.3). Consequently, Fermion doubling appears in both dimensions (azimuthal and longitudinal).

In the following, we visualize the effect of the effective potential $|V_l(s)|$ with a concrete example. We create a nanowire representing a smooth constriction [see upper panel of Fig. 6.2 a)] with the radial profile

$$R(s) = (r_0 - r_1)\left(1 - \mathrm{e}^{-\frac{L^2}{4b^2}}\right)\left(1 - e^{-\frac{(s-L/2)^2}{b^2}}\right) + r_1, \tag{6.8}$$

where b is a parameter determining the smoothness (large b corresponds to strong smoothing), L is the arc length of the nanowire, and r_0, r_1 are the radii in the outer and central regions, respectively. For the computation of the conductance, we choose a set of parameters (given in the caption of Fig. 6.2) and add, as usual, cylindrical leads. In order to study the pristine effect of the effective potential on the transport properties of the curved nanowire, we do not add disorder or highly-doped leads. The effective potential $|V_l(s)|$ and the corresponding conductance G as a function of the Fermi energy ϵ_F can be seen in Fig. 6.2. In Sec. 5.4 where we discussed nanocones in coaxial magnetic fields, we explained in great detail how the position of individual conductance steps can be predicted by an analysis of the effective potential. With the smooth nanowire constriction considered here,

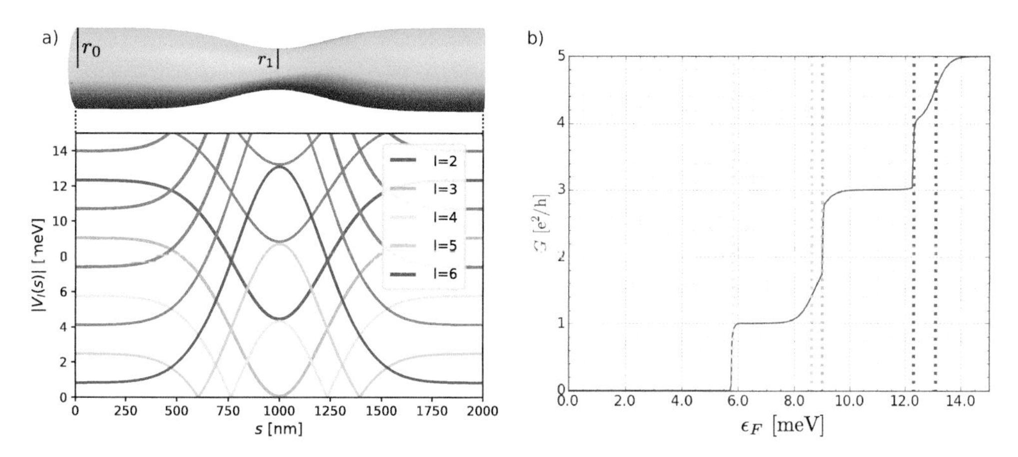

Figure 6.2: a) Smooth nanowire constriction with an outer radius r_0 and an inner radius r_1. The corresponding effective potential for $B = 0.82\,\mathrm{T}$ is shown for all relevant angular momentum quantum numbers l in the given energy range. The magnetic flux at the beginning of the constriction ($s = 0$) is $\Phi = 6.25\,\Phi_0$ and in the center ($s = 1000\,\mathrm{nm}$) $\Phi = 3.06\,\Phi_0$. The former explains the equidistant spacing of $|V_l(s = 0)|$ and the latter explains that the curves in the central region almost touch. The effective potential curves which are responsible for conductance steps in b) are highlighted with colors. b) Conductance through a clean nanowire as sketched in a) as a function of the Fermi energy ϵ_F computed with *kwant*. The maximum energy of each colored effective potential $|V_l(s)|$ in a) is marked with a vertical dashed line [for the color code see the legend in a)]. The outer radius of the nanowire is $r_0 = 100\,\mathrm{nm}$, the inner radius is $r_1 = 75\,\mathrm{nm}$, and the arc length is $L = 2000\,\mathrm{nm}$. In order to avoid Fermion doubling we use a Wilson mass term (see Sec. 2.4.2).

we can extend this discussion and even explain how the effective potential affects the shape of the conductance steps.

We start by observing that transport is blocked below $\epsilon_F \approx 6\,\mathrm{meV}$ because all modes present in the leads cannot pass the potential barriers they face in the center [see, for instance $l = 6$ (purple) and $l = 5$ (blue)].[3] At $\epsilon_F \approx 6\,\mathrm{meV}$, the $l = 4$ mode entering the scattering region can directly pass through since the local maximal value of the effective potential in the center $|V_{l=4}(s = 1000\,\mathrm{nm})|$ is only slightly above $4\,\mathrm{meV}$. Hence we observe a sharp conductance step at

[3]The blocking of modes can be interpreted as follows. Let us, for now, assume that the Fermi energy is at $2\,\mathrm{meV}$ (where conductance is zero). The mode $l = 6$ enters the scattering region from the left lead with an angular momentum $p_\varphi = V_{l=6}(s = 0)/v_F$ and a longitudinal momentum $p_s = \sqrt{\epsilon_F^2/v_F^2 - p_\varphi^2}$. While proceeding into the scattering region, the magnitude of the angular momentum increases since the effective potential $|V_l(s)|$ rises, while the longitudinal momentum decreases. At $s \approx 500$ nm, we have $p_s = 0$ and $p_\varphi = \epsilon_F/v_F$, *i.e.* the electronic motion is purely angular. At this point, it is reflected and returns to the left lead.

$\epsilon_F \approx 6\,\mathrm{meV}$. A different situation occurs for the $l = 5$ mode (blue). It opens already slightly above $\epsilon_F \approx 2\,\mathrm{meV}$ in the lead, but the potential maximum takes the value $|V_{l=5}(s = 1000\,\mathrm{nm})| \approx 8.5\,\mathrm{meV}$. Comparing the conductance step around $\epsilon_F \approx 8.5\,\mathrm{meV}$ with the one at $\epsilon_F \approx 6\,\mathrm{meV}$, it is apparent that its slope is lower. The reason for this is that the electronic mode can tunnel through the $l = 5$ barrier at energies below $8.5\,\mathrm{meV}$, leading to a finite conductance contribution already below $8.5\,\mathrm{meV}$. The same behavior can be observed for the $l = 6$ effective potential (purple curve).

Note that due to the rotational symmetry, coupling between different l-modes is absent and the crossings in Fig. 6.2 a) are real crossings and not avoided crossings. Hence, an electron cannot traverse the wire by changing its angular momentum quantum number. As soon as rotational symmetry is broken by a gate or disorder, coupling between l-modes can change transport through the nanowire drastically. We have seen an example for this in Sec. 5.4, where we discussed transport through a 3DTI nanocone. Here, coupling between l-modes enabled electrons to traverse the nanocone via bound states of the effective potential (which turned out to be QH states) leading to resonant transmission through LLs.

6.2 Coulomb blockade in smoothed topological insulator nanocones

In the last section, we have seen that magnetic barriers induced by the effective potential $|V_l(s)|$ appear when applying a coaxial magnetic field to a curved 3DTI nanowire. An interesting question that can be posed is whether such barriers can be exploited to confine Dirac electrons, a task which cannot be achieved with electrostatic potentials due to Klein tunneling. Confining Dirac electrons in graphene by using inhomogeneous magnetic fields has been proposed in Ref. [128]. However, creating such inhomogeneous magnetic fields on small length scales can be challenging from an experimental point of view. Instead, in curved 3DTI nanowires all that is needed is a homogeneous magnetic field – the spatially varying perpendicular magnetic field component is then realized through the geometry of the nanowire. Utilizing the simple mathematical form of the effective potentials $|V_l(s)|$ [see Eq. (6.7)], geometries leading to all sorts of barrier shapes can be readily envisaged and designed, and transport properties of nanowires can be tailored accordingly. Confining Dirac electrons within barriers of the effective potential brings us to the concept of a *quantum magnetic bottle*, which is a quantum mechanical version of the classical magnetic bottle. The latter traps electrons via the Lorentz force, similar to the Van Allen radiation belt originating from the earth magnetic field [129, 130].

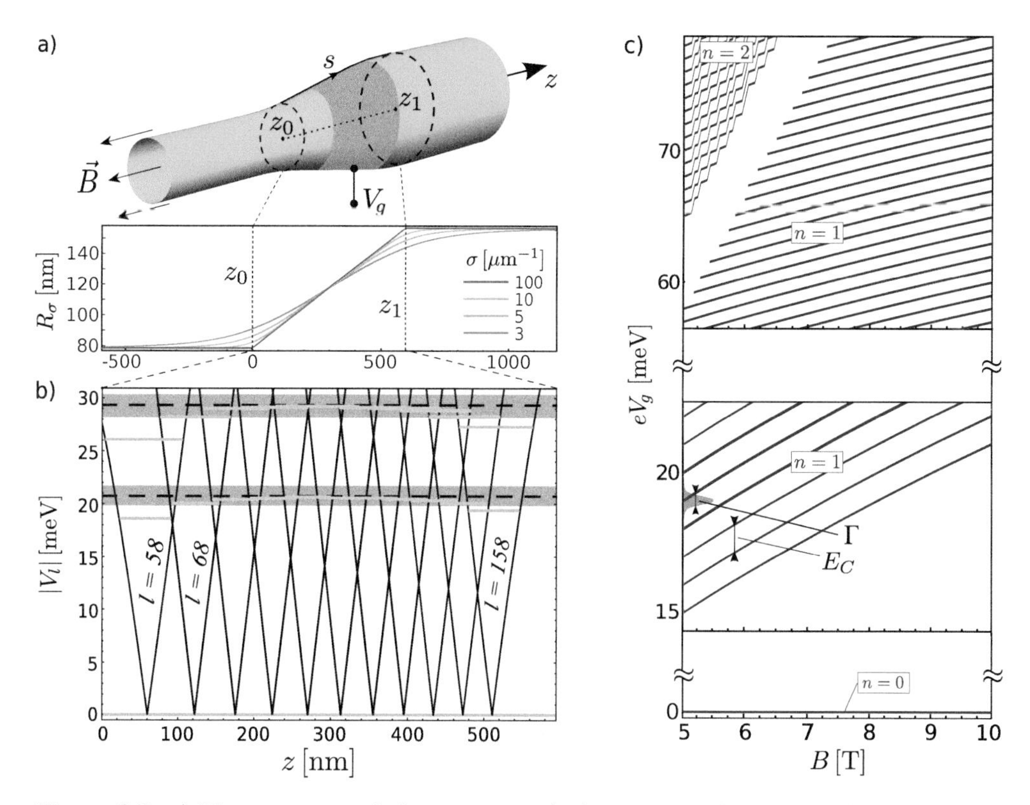

Figure 6.3: a) The upper panel shows a smoothed nanocone characterized by the radial profile $R_\sigma(z)$ with $\sigma = 10\,\mu\text{m}^{-1}$. The radius function $R_\sigma(z)$ is shown in the lower panel for several values of σ. The gray region in the center sketches a gate wrapping around the nanocone, which can be used to tune the electrostatic potential homogeneously. In the literature, this is referred to as "surrounding-gate architecture" or "lateral wrap gate" [131, 132]. b) Corresponding effective potentials $|V_l(s)|$ for $B = 10\,\text{T}$ with bound state energies (orange lines) and LL energies (black dashed horizontal lines) for the sharp version of the cone ($\sigma \to \infty$). Effective potentials $|V_l(s)|$ for angular momentum quantum numbers $l = 58$ to $l = 158$ in step sizes of $\Delta l = 10$ are shown. For $n > 0$, bound (QH) state energies at the borders of the nanocone are lowered substantially compared to the LL energies, while the $n = 0$ bound states are pinned to zero. Gray shade around the LL energies depicts the disorder broadening Γ of the conductance peaks originating from resonant transport through LLs of the nanocone used in Fig. 5.7 a) for a disorder strength of $K = 0.1$. c) Addition spectrum of the inner island QH states that are within the disorder broadening in b) as a function of magnetic field. While the $n = 0$ LL is not affected by interactions and thus yields only a single line, $n > 0$ LLs are split by the charging energy E_C (here we used $E_C = 1\,\text{meV}$). The parameters are chosen such that in the limit $\sigma \to \infty$ the cone studied here agrees with the cone geometry studied in Ch. 5. Adapted from Ref. [124].

As presented in the following, the possibly simplest wire geometry for such a quantum magnetic bottle is a smooth version of the nanocone studied in Ch. 5, which can be seen in Fig. 6.3 a). Here, the kink-like connections to the leads are smoothed, while the radius in the central part of the cone still changes (approximately) linearly. Apart from being a quantum magnetic bottle, a smoothed nanocone is also interesting from an experimental point of view. After all, smooth connections to the leads are much closer to a realistic system than sharp kinks.

6.2.1 Effective potential of the smoothed nanocone

For the description of the smoothed nanocone, we work with the coaxial coordinate z instead of the arc length coordinate s. In order to construct its surface of revolution Σ, we introduce a smoothed version of the Heaviside step function $\Theta_\sigma(z-z') \equiv \frac{1}{2}+\frac{1}{\pi}\arctan[\sigma(z-z')]$ where the strength of the smoothing is determined by σ (small σ corresponds to strong smoothing), and which yields the step function $\Theta(z-z')$ in the limit $\sigma \to \infty$. The radial profile $R_\sigma(z)$ of the smoothed nanocone that starts with a cylinder of radius R_0 at $z=-\infty$ and becomes a cylinder of radius R_1 at $z=\infty$ can then be written as

$$R_\sigma(z) = R_0 + (R_1 - R_0)\Theta_\sigma(z-z_1) + \mathcal{S}(z-z_0)[\Theta_\sigma(z-z_0) - \Theta_\sigma(z-z_1)], \quad (6.9)$$

where $\mathcal{S}$ is the slope of the corresponding sharp version of the nanocone. The lower panel in Fig. 6.3 a) shows $R_\sigma(z)$ for several values of σ.

In the following, we work with the nanocone geometry studied in Ch. 5 with a smoothing parameter of $\sigma = 10\,\mu\mathrm{m}^{-1}$. For reasons that will become apparent shortly, we increase the magnetic field strength to $B = 10\,\mathrm{T}$. The corresponding effective potentials are shown in Fig. 6.3 b) for angular momentum quantum numbers $l = 58$ to $l = 158$ (for visibility we use steps $\Delta l = 10$). The energies of the bound states within each effective potential wedge are obtained by solving the second order differential equation (5.15) with the effective potential Eq. (6.7), and are marked by orange horizontal lines. The black horizontal lines indicate the LL energies for $n = 1$ and $n = 2$ for the sharp cone, *i.e.* for $\sigma \to \infty$. Close to the leads (*i.e.* close to z_0 and z_1), the energies of the bound states with $n > 0$ are substantially lowered compared to the central ones which align well with the LL energies of the sharp cone. The reason for this is that the magnetic field component perpendicular to the surface $B_\perp$ is lowered close to the leads as compared to the central region, which results in a local decrease of the QH state energies [see Eq. (3.24)]. Note that the energy of the LLL (the LL with index $n = 0$) is pinned to zero independently of the perpendicular magnetic field component $B_\perp$, which is a distinct feature of Dirac LLs.

If the lowering of the bound state energies close to the leads is strong enough such that they are no longer within the disorder broadening of the $n > 0$ LLs, they are not in resonance with the central QH states any more. The disorder broadening taken from Fig. 5.7 a) for a disorder strength of $K = 0.1$ is sketched by a gray shade around the LL energies in Fig. 6.3 b). Here, the QH state energies at the borders ($z \approx 60\,\mathrm{nm}$ and $z \approx 510\,\mathrm{nm}$) are beyond the disorder broadening, and thus no longer contribute to resonant transport. We can conclude that magnetic barriers close to the leads arise when the Fermi energy is aligned with the (disorder broadened) LL energy of the central region of the cone. Hence, a QH island separated from the leads by magnetic barriers emerges.

The condition for this scenario is that QH states form in the smoothed border region of arc length l with considerably lowered slope, where the local perpendicular magnetic field component $B_\perp$ leads to QH state energies below the disorder broadening. This is only possible if the local magnetic length l_B is smaller than the size of this region, *i.e.* $l_B \ll l$. The cone parameters used here together with a magnetic field strength of $B = 2\,\mathrm{T}$, a disorder strength of $K = 0.1$, and a smoothing of $\sigma = 10\,\mu\mathrm{m}^{-1}$, do not fulfill this requirement (with these parameters the outermost QH states are within the disorder broadening). Hence, we increase the magnetic field to $B = 10\,\mathrm{T}$, which lowers the magnetic length such that $l_B \ll l$. Alternatively, one could increase l by enlarging the overall size of the nanowire or by making it smoother (*i.e.* by decreasing σ). Another alternative is to reduce disorder broadening by reducing the disorder strength K. Note that if the size of the smoothed region where $\epsilon_n = v_F\sqrt{2\hbar e B_\perp n}$ for $n > 0$ is below the disorder broadening is smaller than the magnetic length, *i.e.* $l < l_B$, transport through the smoothed nanocone is determined by resonant transmission through LLs, as described in Sec. 5.

6.2.2 Addition spectrum

Since the magnetic barriers induced by the effective potentials lead to an isolated island of QH states in the central region of the smoothed cone for LLs with $n > 0$, Coulomb blockade should be observable when varying the gate voltage V_g of an electrode applied to the central region. An exact description of Coulomb blockade physics of QDs in strong magnetic fields is, in general, a complicated many-body problem. However, general features of the addition spectrum can usually be obtained from the so-called constant interaction model, which is based on a constant charging energy $E_C = e^2/(2C)$, and the single-particle spectrum of the QD [42]. Since we are only after general features and do not aim at a quantitative description which depends on details of the experimental setup, we will make only use of the constant interaction model introduced in Sec. 2.1.

The Coulomb split QH state spectrum for $n = 1$ and $n = 2$, and the unaffected LLL are shown in Fig. 6.3 c) as a function of the magnetic field. Each LL with $n > 0$ composed of N quasi-degenerate QH states is split into N lines, where the size of the splitting is determined by the charging energy E_C. The lower right set of QH states belong to the $n = 1$ LL, the upper left set to the $n = 2$ LL. In order to observe Coulomb blockade, E_C has to be larger than the disorder broadening Γ (sketched with gray shade). The lines of the $n = 1$ LL move upward with increasing magnetic field since the LL energies behave as $\epsilon_n \propto \sqrt{B_\perp}$ [see Eq. (3.24)]. Moreover, the degeneracy N of each LL increases with increasing B since more QH states fit into the central region (because of $l_B \propto 1/\sqrt{B_\perp}$). Hence, additional lines appear with increasing magnetic field, which is visible in the upper half of Fig. 6.3 c). The gap between the $n = 1$ and $n = 2$ LL for fixed B is the Landau gap $\epsilon_2 - \epsilon_1 = v_F\sqrt{2\hbar e|B_\perp|}$. Each time the degeneracy of the $n = 1$ LL increases by one the bottom of the $n = 2$ LL is shifted upwards by E_C, leading to the step-like structure in the upper left corner.

Notably, there are clear signatures of the Dirac nature of our system: the single-particle energies are proportional to $\sqrt{B}$ as opposed to trivial electrons whose energies are linear in B (in the QH phase). Most prominently is the fact that the LLL is pinned to zero energy and thus transport stays resonant resulting in a single line in Fig. 6.3 c).

6.2.3 Experimental realization and concluding remarks

Smoothed nanocones, as proposed above, are within experimental reach: for instance, HgTe-based 3DTI nanowires [28] and core-shell nanowires [119] with spatially varying cross sections have been built for transport experiments. Moreover, our conclusions do not require the nanocone to have a truly circular cross section. Deformations lift the degeneracies of the QH states within each LL, but as long as this splitting is smaller then the broadening due to, *e.g.* temperature or disorder, no qualitative change in the conductance is expected.

The charging energy E_C depends on the setup size, geometry, and materials including the dielectrics, and can thus be tuned in a broad range. The charging energy E_C of the nanowire based on strained HgTe as studied in Ch. 4 and Ref. [28] is of the order of 1 meV, which is the value we used in Fig. 6.3 c).

The strength of the tunnel coupling between the leads and the inner island QH system depends on the size of the magnetic barriers in the smoothed border regions, which are determined by the shape of the effective potentials $|V_l(s)|$. Tunneling from the leads into the central QH island thus depends on the precise geometry of the nanowire, as well as the magnetic field and the disorder strength.

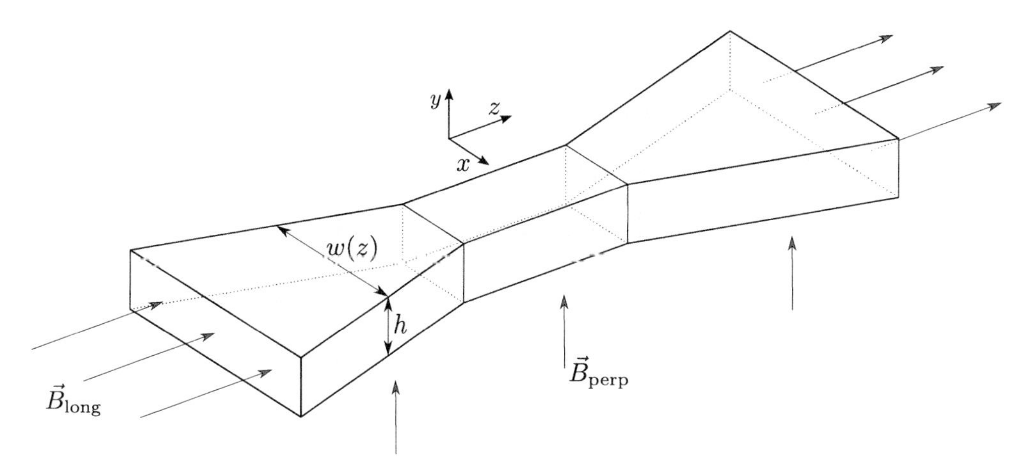

Figure 6.4: Shaped nanowire with width profile $w(z)$ and constant height h in a longitudinal and perpendicular magnetic field $\mathbf{B}_{\text{long}}$, $\mathbf{B}_{\text{perp}}$.

A computation of the tunneling rates is quite involved and beyond the scope of this thesis.

Finally, note that many other geometries can be envisaged to act as a quantum magnetic bottle, as for example the curved nanowire shown in Fig. 6.1. Other wire geometries acting as possible quantum magnetic bottles are discussed in Ref. [133].

6.3 Nanowire constriction with rectangular cross section

Up to this point, we have only considered TI nanowires with rotational symmetry, where the Hamiltonian takes the simple form of Eq. (6.4). However, in this section we will study an example of a nanowire with broken rotational symmetry, namely a nanowire with rectangular shape where the height h is constant while the width $w(z)$ varies along the wire direction (see Fig. 6.4). The motivation to study such a geometry stems from the experimental side: The systems hosting strained HgTe nanowires studied in Ch. 4 and Ref. [28] consist of a layered structure as shown in Fig. 4.1 a), where the HgTe layer has, for instance, a constant height of 80 nm. With an etching process the nanowire is "cut out" from the top [see Fig. 4.1 b)]. If the variation of the width $w(z)$ is negligible, the cross section along the wire direction is approximately constant and the rectangular nanowire can be described by Eq. (3.2). However, as soon as the cross section changes along the wire direction, its precise shape can have a large effect on the transport properties.

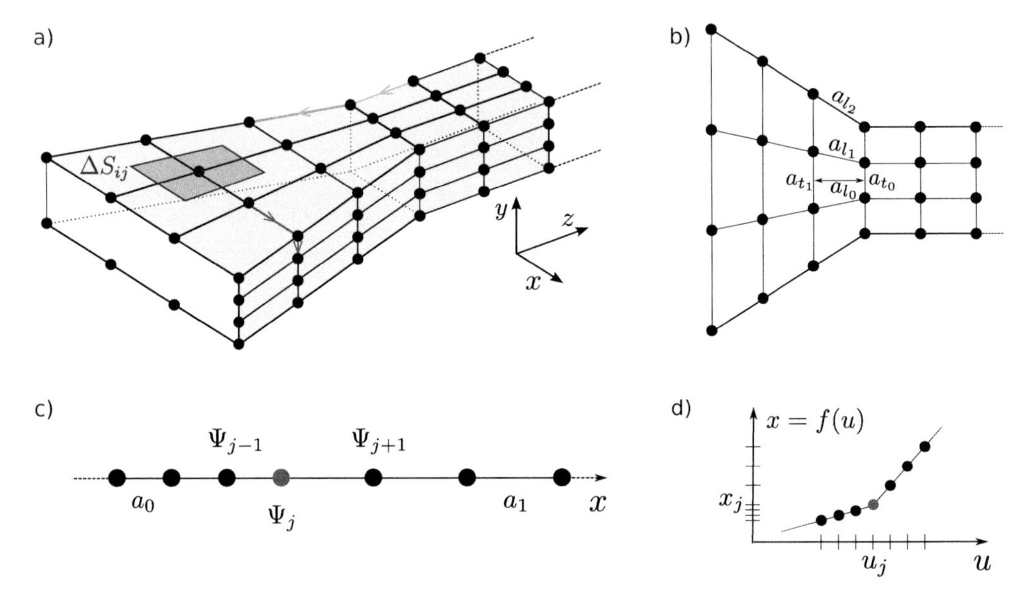

Figure 6.5: a) Part of the shaped nanowire shown in Fig. 6.4 with the numerical grid we use for simulations. In general, the hopping amplitude changes along the hopping direction (see, for instance, the green or red path). b) Top view of the numerical grid. The grid spacings can be computed using Eq. (6.19). c) 1D chain with an abrupt change in the lattice constant (from a_0 to a_1) at site j. d) Coordinate transformation $x = f(u)$ describing the non-uniform grid depicted in c).

As an example, we choose the wire sketched in Fig. 6.4: two taperings with linearly changing width $w(z) = az$ connected by a wire with constant width $w(z) = w_{\text{center}}$. The height h is kept constant throughout the entire nanowire and the cross section is rectangular. This geometry is within experimental reach, since the etching process mentioned above can be used to create almost any width profile $w(z)$ (even Aharonov-Bohm rings have been produced). For the rest of this chapter, we will study such nanowires in magnetic fields numerically. A crucial ingredient for the simulations is a non-uniform numerical grid, introduced in the following.

6.3.1 Non-uniform grid

In order to simulate magnetotransport though the TI nanowire depicted in Fig. 6.4, we construct a tight-binding Dirac Hamiltonian on the numerical grid sketched in Fig. 6.5 a) (side view) and Fig. 6.5 b) (top view). Here, the number of grid points in the transversal direction (around the perimeter) is kept constant, while the grid spacings $a(\mathbf{r}_{ij}, \mathbf{r}_{i'j'})$ between two neighboring grid points located at $\mathbf{r}_{ij}$ and $\mathbf{r}_{i'j'}$ in general vary. There is a fundamental difference between the grid used here compared to the grid we use for nanowires with rotational symmetry, as introduced, for instance, in Sec. 5.3. For the latter, we adopt a constant longitudinal grid spacing a_s while the transversal grid spacing $a_\varphi(s)$ is a function of the longitudinal arc length coordinate s. It is important to note that for such systems the grid spacing and consequently the hopping integral never change *along* the direction of the hopping. For the grid depicted in Fig. 6.5 a), this is no longer true. Clearly, the hoppings along the red and green paths in Fig. 6.5 a) involve an abrupt change of the grid spacing.

In order to understand the consequences of this abrupt change let us consider a 1D chain labeled by a grid index $i \in \mathbb{Z}$ where the grid spacing changes at $i = j$ at the position $x = x_j$ from a_0 to a_1 as shown in Fig. 6.5 c). Acting with the operator $\hat{k} = -\mathrm{i}\partial_x$ on a function $\Psi(x)$ on this grid at positions x_{j-1}, x_j, and x_{j+1} with a finite difference scheme, yields

$$\hat{k}\Psi(x)\Big|_{x=x_{j-1}} = -\mathrm{i}\frac{\Psi_j - \Psi_{j-2}}{2a_0}, \tag{6.10}$$

$$\hat{k}\Psi(x)\Big|_{x=x_j} = -\mathrm{i}\frac{\Psi_{j+1} - \Psi_{j-1}}{a_0 + a_1}, \quad \text{and} \tag{6.11}$$

$$\hat{k}\Psi(x)\Big|_{x=x_{j+1}} = -\mathrm{i}\frac{\Psi_{j+2} - \Psi_j}{2a_1}. \tag{6.12}$$

Using the above equations to construct a tight-binding version of the 1D Dirac Hamiltonian $H = v_F\hat{k}\sigma_x$ as shown in Sec. 2.4.1, it turns out that H on this grid is not Hermitian since the hopping $j \to j+1$ involves the term $-\mathrm{i}/(a_0 + a_1)$ while the hopping $j+1 \to j$ involves $\mathrm{i}/(2a_1)$ (for H to be Hermitian they have to be equal up to complex conjugation). However, as explained in Sec. 5.1, a Hermitian tight-binding matrix is necessary for simulations with *kwant*.

In order to resolve this problem, we express the varying grid spacing in terms of a coordinate transformation $x = f(u)$, where u is a continuous version of the grid index $i \in \mathbb{Z}$, *i.e.* it is dimensionless. The $u_i \equiv i$ values are evenly spaced and form the so-called *logical* grid with $x_i = f(u_i)$. In our simple example, the transformation takes the form

$$f(u) = \begin{cases} a_0 u & u \le u_j \\ a_1 u & u > u_j \end{cases} \tag{6.13}$$

which is depicted in Fig. 6.5 d). The Hamiltonian H can be expressed in the u-coordinate, yielding

$$\tilde{H} = -\mathrm{i}v_F\sigma_x\frac{\partial u}{\partial x}\frac{\partial}{\partial u} = -\mathrm{i}v_F\sigma_x\frac{1}{f'(u)}\partial_u \tag{6.14}$$

with the corresponding eigenfunctions $\tilde{\Psi}(u) = \Psi(f(u))$. One can easily show that a finite difference scheme on the logical grid applied to Eq. (6.14) gives the same non-Hermitian tight-binding matrix as obtained by using Eqs. (6.10) - (6.12). However, we are now able to argue that the reason for the non-Hermiticity of the tight-binding matrix is the non-trivial volume form $\mathrm{d}V = f'(u)\mathrm{d}u$ induced by $f(u)$. Hence, we can use the transformation given by Eq. (5.4) which renders the volume form trivial and thus results in a Hamiltonian $\bar{H}$ with a Hermitian tight-binding form (see Sec. 5.1). The transformed Hamiltonian $\bar{H}$ reads

$$\bar{H} \equiv \sqrt{f'(u)}\tilde{H}\frac{1}{\sqrt{f'(u)}} = -\mathrm{i}v_F\sigma_x\frac{1}{\sqrt{f'(u)}}\partial_u\frac{1}{\sqrt{f'(u)}}, \tag{6.15}$$

and the corresponding eigenfunctions are given by $\bar{\Psi}(u) = \sqrt{f'(u)}\tilde{\Psi}(u)$. For convenience, we denote $p(u) \equiv 1/\sqrt{f'(u)}$ in the following and use the notation $p_j \equiv p(u_j)$ on the logical grid. The action of $\bar{H}$ on a wave function $\bar{\Psi}(u)$ on the grid is then described by

$$-\mathrm{i}p(u)\partial_u p(u)\Psi(u)\Big|_{u=u_{j-1}} = -\frac{\mathrm{i}}{2}p_{j-1}\left(p_j\Psi_j - p_{j-2}\Psi_{j-2}\right) \tag{6.16}$$

$$-\mathrm{i}p(u)\partial_u p(u)\Psi(u)\Big|_{u=u_j} = -\frac{\mathrm{i}}{2}p_j\left(p_{j+1}\Psi_{j+1} - p_{j-1}\Psi_{j-1}\right) \tag{6.17}$$

$$-\mathrm{i}p(u)\partial_u p(u)\Psi(u)\Big|_{u=u_{j+1}} = -\frac{\mathrm{i}}{2}p_{j+1}\left(p_{j+2}\Psi_{j+2} - p_j\Psi_j\right). \tag{6.18}$$

From the second line in the above equations, we can extract that the hopping $j \to j+1$ is given by the term $-\frac{\mathrm{i}}{2}p_jp_{j+1}$, while the hopping $j+1 \to j$, which can be extracted from the third line, is given by the corresponding complex conjugate $\frac{\mathrm{i}}{2}p_jp_{j+1}$. Consequently, the tight-binding matrix form of $\bar{H}$ is Hermitian.

The above results for the 1D chain can be generalized to the grid shown in Fig. 6.5 a) by constructing the functions $f(u)$ for the transversal and longitudinal direction. One helpful relation in this regard is

$$a_{l_n} = \sqrt{a_{l_0}^2 + (2n-1)^2(a_{t_1} - a_{t_0})^2/4} \quad \text{for} \quad n = 1, 2, 3, ..., \tag{6.19}$$

where the longitudinal grid spacings a_{l_n} for $n = 0, 1, 2$ as well as the transversal grid spacings a_{t_n} for $n = 0, 1$ are depicted in Fig. 6.5 b). In order to compute the physical wave functions $\Psi(x, y)$, the volume form transformation, Eq. (6.15), has to be taken into account, which can be done by rescaling the numerically obtained wave functions according to $\Psi(x_i, y_j) = \bar{\Psi}_{ij}/\Delta S_{ij}$, where ΔS_{ij} is the Wigner-Seitz cell around the grid point $\mathbf{r}_{ij}$ which is sketched in blue in Fig. 6.5 a).

Note that the application of the non-uniform grid described above is not restricted to Dirac Hamiltonians, and that other grid geometries can be implemented analogously. This might be useful especially when simulating full 3D bulk systems, which is numerically very costly. Using the technique described above, it is possible to use more grid points in regions of particular interest. For instance, when simulating a 3DTI with a bulk model, only the surface states might be of interest, and hence the number of grid points in the bulk can be reduced saving computational capacity.

Magnetic fields

We implement magnetic fields using Peierls substitiution, which was discussed in Sec. 3.3. Here, a phase factor $\alpha_\gamma(\boldsymbol{A}) = \exp\left(-\mathrm{i}\frac{e}{\hbar}\int_\gamma \mathrm{d}\boldsymbol{l}\,\boldsymbol{A}(l)\right)$ is added to the hopping along a path γ in a vector potential $\boldsymbol{A}$.

For the constant-height wire subject to a longitudinal magnetic field, a convenient gauge choice for the vector potential is given by $\boldsymbol{A}_{\text{long}} = B_{\text{long}}\,(0, x, 0)^T$ in the Cartesian coordinate system depicted in Fig. 6.5 a). Since $\boldsymbol{A}_{\text{long}}$ has only non-zero components in the y-direction, only the hoppings on the sides of the nanowire along the y-direction acquire a non-trivial phase factor, which is given by

$$\alpha_\gamma(\boldsymbol{A}_{\text{long}}) = \exp[-\mathrm{i}2\pi B_{\text{long}} a_y W(z)/(2\Phi_0)]\,, \tag{6.20}$$

where a_y is the corresponding grid spacing.

For the perpendicular magnetic field, it is convenient to use the vector potential $\boldsymbol{A}_{\text{perp}} = B_{\text{perp}}\,(0, 0, -x)^T$ which only gives phase contributions to hoppings along the z-direction. Note, however, that the x-component is in general not constant during longitudinal hopping [see Fig. 6.5 a)]. In order to account for that, we parametrize the path along a longitudinal hopping γ_l starting at a grid point $\mathbf{r} = (x, y, z)^T$ and ending at $\mathbf{r} = (x + \Delta\, x, y, z + \Delta\, z)^T$ as

$$\gamma_l : [0, 1] \to \mathbb{R}^3, \quad t \to (x + \Delta x\, t, y, z + \Delta z\, t), \tag{6.21}$$

which allows to compute the integral

$$\int_{\gamma_l} \mathrm{d}\boldsymbol{l}\boldsymbol{A} = \int_0^1 \mathrm{d}t\boldsymbol{A}[\gamma_l(t)] \cdot \frac{\mathrm{d}}{\mathrm{d}t}\gamma_l(t) = \int_0^1 \mathrm{d}t \begin{pmatrix} 0 \\ 0 \\ -B(x + \Delta x\, t) \end{pmatrix} \cdot \begin{pmatrix} \Delta x \\ 0 \\ \Delta z \end{pmatrix} \tag{6.22}$$

$$= -B\Delta z(x + \Delta x/2). \tag{6.23}$$

The corresponding phase in the longitudinal hopping terms is then given by

$$\alpha_{\gamma_l}(\boldsymbol{A}_{\text{perp}}) = \exp[\mathrm{i}2\pi\Delta z B(x + \Delta x/2)/\Phi_0]\,. \tag{6.24}$$

As expected, it is the average position $x + \Delta x/2$ which enters the phase factor in the hopping.

6.3.2 Magnetotransport simulations

In the following, we study the magnetotransport properties of the nanowire constriction with rectangular cross section, sketched in Fig. 6.4, by means of numerical simulations with *kwant*. To this end, we use a tight-binding model based on the numerical grid which was introduced in the last section to which we add semi-infinite rectangular leads. Moreover, we add Gaussian correlated disorder created in a 3D box with the FFM as described in Secs. 3.2.1 and 5.3.

Perpendicular magnetic field

We start by considering the nanowire in a perpendicular magnetic field with strength $B_{\text{perp}} = 2\,\text{T}$ ($l_B \approx 18.1\,\text{nm}$) as sketched in Fig. 6.6 c) with a height of $h = 80\,\text{nm}$, maximal width (at the connection to the leads) of $w_{\text{lead}} = 150\,\text{nm}$, and central width $w_{\text{center}} = 75\,\text{nm}$. Moreover, we choose $L_{\text{center}} = 100\,\text{nm}$ for the length of the central region and $L_{\text{arc}} = 100\,\text{nm}$ for the outer arc length between central region and leads. It is instructive to start by studying the band structure of the leads in perpendicular magnetic field, which is shown in Fig. 6.6 a). The LL energies given by Eq. (3.24) with $B = 2\,\text{T}$ are marked by horizontal dashed lines. Flat bands form at these energies in a large k-range due to QH state formation on the top and bottom surface. Note that there is a difference between the band structure we consider her, and the band structure of cylindrical nanowires in perpendicular magnetic field shown in Fig. 3.9 a) and Fig. 3.11 b). While the former is completely flat up to a k-value for which the corresponding QH state resides on one of the corners of the nanowire, the latter shows a rounded shape, *i.e.* it disperses for all k-values. The reason for this is that for the rectangular cross section the perpendicular magnetic field component piercing the surface changes abruptly from 2 T to 0 T at the corner of the nanowire, while for a cylindrical nanowire it changes gradually with $B\cos(\varphi)$. For larger k-values, the absolute value of the energy first decreases and then increases again, which is a remainder of the 1D Dirac cone residing on the sides of the wire where the perpendicular magnetic field component is zero [see Fig. 3.9 a)]. We will see that those highly dispersing states are chiral side surface states which dominate transport.

Fig. 6.6 b) shows the conductance through the disordered nanowire as a function of the Fermi energy for a disorder strength of $K = 0.2$. Note that the conductance shown here corresponds to one single disorder configuration, which explains the appearance of large fluctuations in the form of local spikes. Nevertheless, we see a disorder robust conductance plateau in the energy range between the LLL and the $n = 1$ LL (up to disorder broadening). This conductance plateau originates from chiral side surface states with opposite chirality on opposite sides, sketched in Fig. 6.6 c) with orange arrows, which are decoupled due to the gapped top and

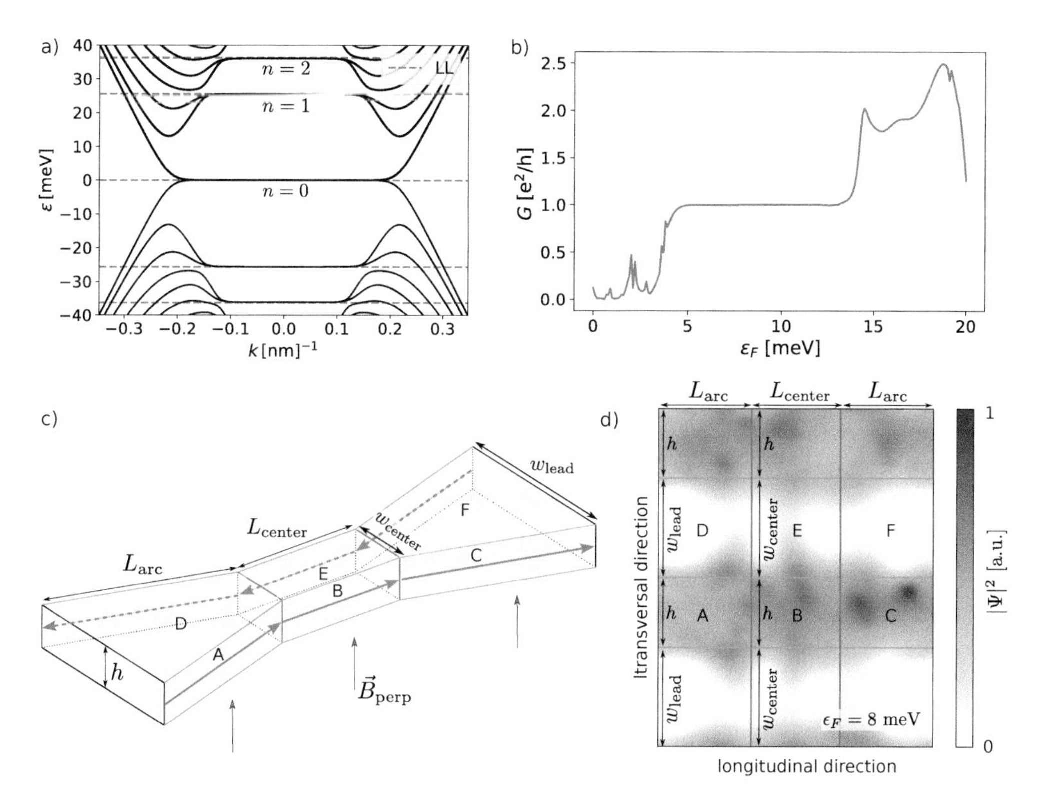

Figure 6.6: a) Band structure of the leads in a perpendicular magnetic field with strength $B_{\rm perp} = 2\,{\rm T}$, calculated with an infinitely long nanowire with a rectangular cross section of height $h = 80\,{\rm nm}$ and width $w_{\rm lead} = 150\,{\rm nm}$. b) Conductance G as a function of the Fermi energy ϵ_F for one single disorder configuration with a disorder strength of $K = 0.2$, and a correlation length of $\xi = 7\,{\rm nm}$. The parameters defining the geometry of the nanowire are given in the main text. Here, we circumvent Fermion doubling by dividing the conductance by four. c) Nanowire with chiral side surface states sketched with orange arrows. d) Probability distribution of the scattering states at $\epsilon_F = 8\,{\rm meV}$ plotted on the logical grid of the unfolded wire. The green and blue lines correspond to the edges colored in c). Matching surfaces in c) and d) are labeled with capital letters.

bottom surfaces which are in the QH phase. The QH phase on top and bottom survives throughout the entire wire because the magnetic length $l_B \approx 18.1\,\text{nm}$ is smaller than $w_{\text{central}} = 75\,\text{nm}$. This observation is corroborated by Fig. 6.6 d), which shows the surface probability distribution $|\Psi(x, y, z)|^2$ of the two counter propagating scattering states that exist at $\epsilon_F = 8\,\text{meV}$ (which lies within the conductance plateau). For plotting we use the logical grid of the unfolded wire, which means that the length scales in Fig. 6.6 d) are not physical. As a guide to the eye, matching surfaces are labeled with capital letters in Fig. 6.6 c) and Fig. 6.6 d). Moreover, edges colored in green and blue in Fig. 6.6 c) correspond to the green and blue lines in Fig. 6.6 d). We can conclude that the probability distribution of the scattering states has indeed a large weight on the side surfaces, and almost vanishes on the top and bottom surfaces of the nanowire.

Oblique magnetic field

In this section we consider an oblique magnetic field $\boldsymbol{B} = \boldsymbol{B}_{\text{long}} + \boldsymbol{B}_{\text{perp}}$ with $B_{\text{long}} = 6\,\text{T}$ and $B_{\text{perp}} = 2\,\text{T}$. From now on, we denote the central region Σ_C and the outer region, *i.e.* the tapering which connects the central region to the leads, Σ_O [see Fig. 6.7 c)]. We slightly change the geometry by increasing the length of Σ_O to $L_{\text{arc}} = 200\,\text{nm}$ while keeping the rest of the parameters unchanged. The oblique magnetic field has a nonzero out-of-plane component $\boldsymbol{B} \cdot \hat{n}_\perp = \boldsymbol{B}_{\text{long}} \cdot \hat{n}_\perp$ on each facet of Σ_O with surface normal $\hat{n}_\perp$. If this out-of-plane component is large enough, *i.e.* if the corresponding magnetic length is small compared to the sizes of the facets, the entire Σ_O surface can be brought into a QH phase, which is the case for the parameters used here (on the sides the magnetic length is $l_B = 21\,\text{nm}$ and on top/bottom we have $l_B = 18.1\,\text{nm}$). In the central region Σ_C, only top and bottom facets are in the QH phase since $\boldsymbol{B}$ does not have out-of-plane components on the side facets.

The band structure of the leads in oblique magnetic field is independent of the longitudinal magnetic field B_{long} as long as the perpendicular magnetic field B_{perp} is strong enough such that top and bottom surfaces are in QH phases. The reason for this is that the QH states on top and bottom do not wrap around the perimeter and consequently do not feel the magnetic flux induced by B_{long}. At large energies, however, the QH phase vanishes and the band structure of a cylinder is recovered [see Fig. 3.9 a)] since the extension of QH states with large LL index n exceeds the width of the wire. However, within the energy range $|\epsilon| < 40\,\text{meV}$, a magnetic field of $B_{\text{perp}} = 2\,\text{T}$, and a width of $w_{\text{lead}} = 150\,\text{nm}$, top and bottom surfaces are well within the QH phase. Consequently, the band structure of the leads in oblique magnetic field (not shown) is almost identical to the band structure shown in Fig. 6.6 a), where $B_{\text{perp}} = 2\,\text{T}$ and $B_{\text{long}} = 0\,\text{T}$.

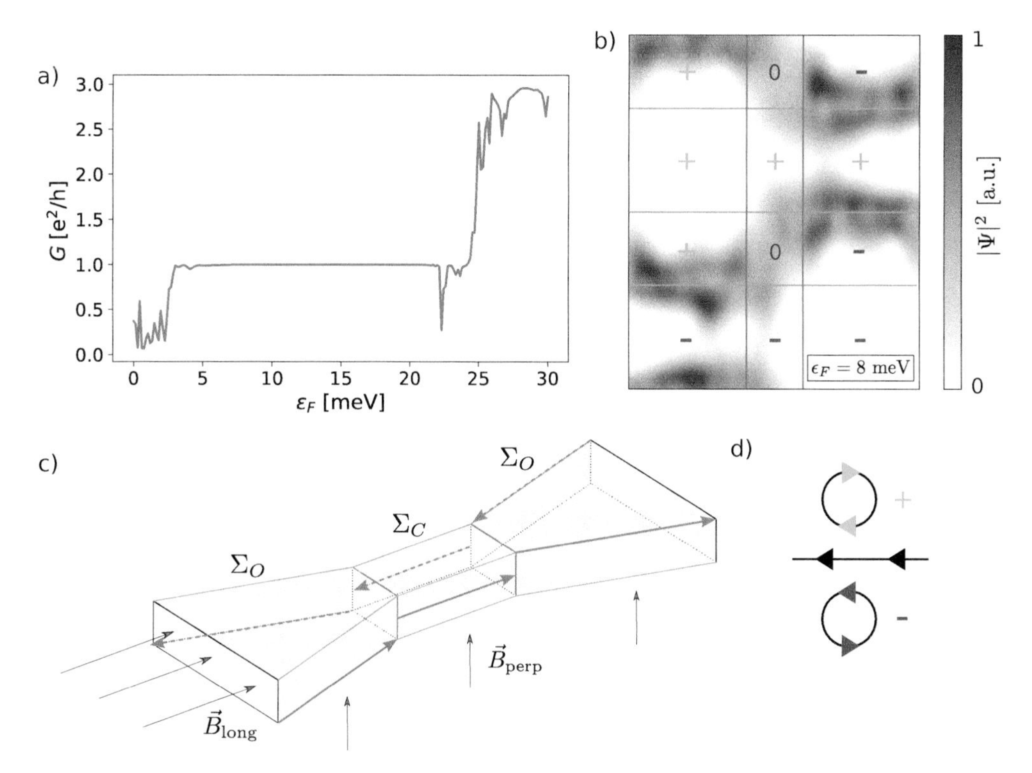

Figure 6.7: a) Conductance G as a function of the Fermi energy ϵ_F for one disorder configuration with a disorder strength of $K = 0.1$, and a correlation length of $\xi = 7\,\text{nm}$. We accounted for Fermion doubling by dividing the conductance by four. b) Probability distribution of the scattering states at $\epsilon_F = 8\,\text{meV}$ plotted on the logical grid of the unfolded wire. For a guide on how to map different regions to the wire surface see Fig. 6.6 c) and d). The sign of the perpendicular magnetic field component $\text{B} \cdot \hat{n}_\perp$ is added to each facet. Increasing L_{arc} to 200 nm (instead of using 100 nm as in Fig. 6.6) highlights the hinge character of the chiral states in region Σ_O. c) Nanowire with 2nd order topological hinge states sketched with orange arrows. d) 1D chrial channel originating from cyclotron orbits with opposite handedness.

Since all side facets of Σ_O are in the QH phase, *i.e.* gapped, one naively expects a qualitative change in the conductance compared to the case $B_{\text{perp}} = 2\,\text{T}$, $B_{\text{long}} = 0\,\text{T}$, where transport is dominated by chiral side surface states. Surprisingly, however, there is no qualitative change of the conductance visible in Fig. 6.7 a) whatsoever. Similar to Fig. 6.6 b), there is a conductance plateau in a large energy range signaling topologically protected surface transport. In order to explain this result, it is instructive to consider the probability distribution of the scattering states at $\epsilon_F = 8\,\text{meV}$ shown in Fig. 6.7 b). We observe that the chiral side surface states from Fig. 6.6 d) in region Σ_O move towards the edges/hinges of the nanowire. This behavior can be explained by higher-order topology: The oblique magnetic field induces an extrinsic second-order TI phase [29–32] in the nanowire which hosts topologically protected chiral hinge states at edges where the sign of the perpendicular magnetic field component $\text{sgn}(\boldsymbol{B} \cdot \hat{n}_\perp)$ changes. In Fig. 6.7 b), the signs of $\boldsymbol{B} \cdot \hat{n}_\perp$ on all facets of the nanowire are shown, and indeed, the probability distribution of the scattering states has almost exclusively weight in areas where the sign changes. The corresponding path of those states is sketched in Fig. 6.6 c) with orange lines. An alternative explanation can be given by semiclassical arguments: Depending on the sign of $\boldsymbol{B} \cdot \hat{n}_\perp$, cyclotron orbits run clockwise or anticlockwise. If domains with different signs meet, cyclotron orbits of different handedness lead to a 1D chiral channel, which is depicted in Fig. 6.7 d).

Longitudinal magnetic field

In the following, we consider the nanowire geometry depicted in Fig. 6.4 in a longitudinal magnetic field $\boldsymbol{B}_{\text{long}}$, which has only out of plane components on the side facets of Σ_O. It turns out that magnetotransport is rather complicated in this case and thus $G(\epsilon_F)$ curves for fixed magnetic fields are no longer fit to explain the occuring physical phenomena. Hence, we resort to a $G(\epsilon_F, B_{\text{long}})$ color map shown in Fig. 6.8 a) for a clean nanowire whose parameters defining its geometry are given in the caption. Before discussing $G(\epsilon_F, B)$, it is instructive to briefly repeat the main magnetoconductance features of nanowires with constant cross section. As mentioned several times in the course of this thesis, such nanowires can be qualitatively modeled by assuming a cylindrical shape. The dispersion of an infinitely long cylindrical nanowire consists of 1D subbands, each corresponding to a different angular momentum, as shown in Fig. 3.2. We have seen in Ch. 3 and Ch. 4 that for clean (dirty) nanowires steps (local minima) in the conductance can be attributed to an alignment of Fermi energy and subband minima, whose energies are given by $\epsilon_{k_z=0,l} = \pm\hbar v_F 2\pi/P|l + 0.5 - \Phi/\Phi_0|$ [see Eq. (3.13)], where $l \in \mathbb{Z}$ is the orbital angular momentum quantum number and P the perimeter of the wire. The magnetic flux $\Phi = BA$ (with cross section A) versus energy ϵ of those subband minima is sketched in Fig. 6.8 d). Depending on the sign of the

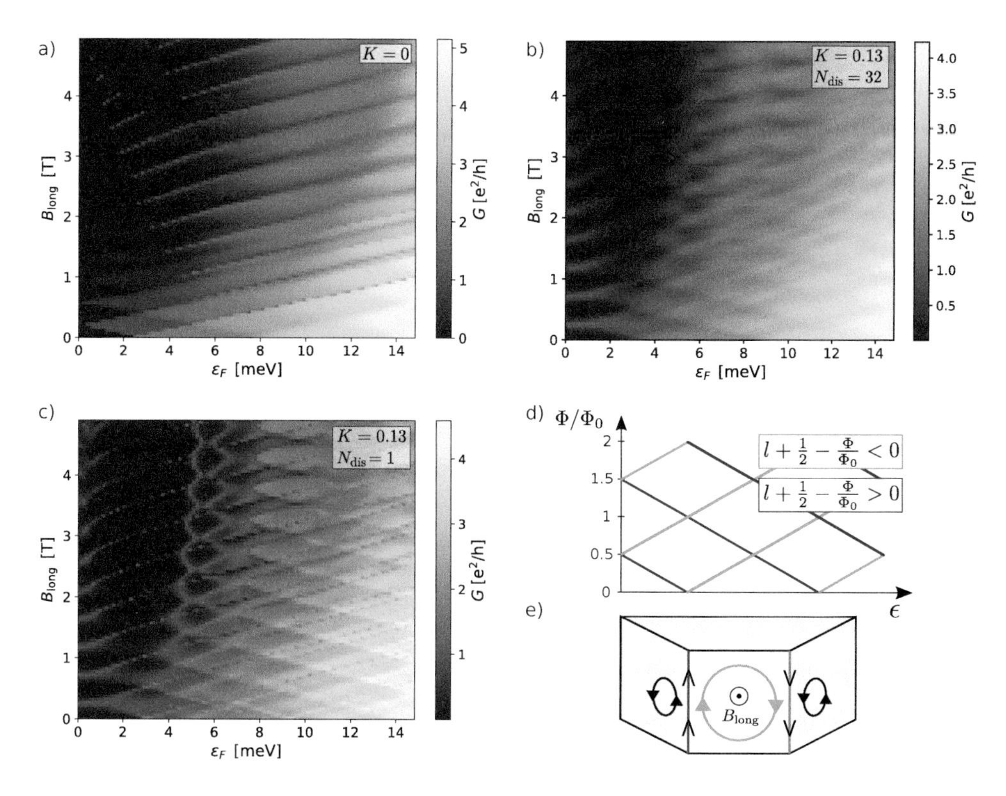

Figure 6.8: a) Color map of the conductance G for a clean nanowire ($K = 0$) as a function of Fermi energy ϵ_F and longitudinal magnetic field $\boldsymbol{B}_{\text{long}}$ for the nanowire geometry shown in Fig. 6.6 c). The parameters are $w_{\text{lead}} = 140\,\text{nm}$, $w_{\text{center}} = 100\,\text{nm}$, $h = 80\,\text{nm}$, $L_{\text{arc}} = 150\,\text{nm}$, and $L_{\text{center}} = 150\,\text{nm}$. Interestingly, lines with only positive slope appear for large magnetic fields, and the slope of these lines shows a kink (for instance around $\epsilon_F \approx 5\,\text{meV}$ and $B_{\text{long}} \approx 3.5\,\text{T}$). Moreover, dark "fingers" (regions with vanishing conductance) intercepting these lines can be observed. b) Disorder-averaged conductance $G(\epsilon_F, B_{\text{long}})$ with disorder strength $K = 0.13$, correlation length $\xi = 15$, and $N_{\text{dis}} = 32$ disorder configurations. The clear lines discussed in a) are washed out. c) $G(\epsilon_F, B_{\text{long}})$ for $K = 0.13$ and $\xi = 15$ for a single disorder configuration. Interesting new conductance features emerge. d) Sketch of the energies of the subband minima of an infinitely long nanowire with constant cross section with a magnetic flux Φ threaded along its axes. e) Front view of Σ_O with cyclotron orbits on the side facets inducing angular motion around the perimeter.

angular momentum $\mathrm{sgn}(l + 0.5 - \Phi/\Phi_0)$, subband minima go up in energy (green lines) or down (blue lines) when increasing the magnetic field. Note that we have seen such a diamond shaped structure already in Fig. 4.7 b), where ϵ_F was tuned by a gate electrode with gate voltage V_g.

For small magnetic fields there is a remainder of those diamonds in Fig. 6.8 a), while for larger magnetic fields lines with positive slope prevail. Interestingly, lines with negative slope are missing, suggesting that only one half of the angular momentum states with the same sign dominate transport. This peculiar behavior suggests that such a nanowire can be used as an angular momentum filter, and, due to spin-momentum locking, even as a spin filter. One possible intuitive explanation for this phenomenon is sketched in Fig. 6.8 e): Cyclotron orbits on the side surfaces of Σ_O lead to either clockwise or anticlockwise motion of electrons around the perimeter at the interface between Σ_O and Σ_C, depending on the sign of the magnetic field. Since this motion has the same sign on both interfaces left and right of Σ_C, this could explain why angular momentum states with only one handedness are transmitted. There are, however, two problems concerning this explanation. The first one is that this kind of angular motion is also induced at the interface to the leads – but with reversed direction. Second, as can be seen in Fig. 6.8 b), using a disordered wire and averaging over several disorder configurations strongly blurs the effect, making it hardly visibly.

Additional interesting effects, especially visible in the clean case shown in Fig. 6.8 a), are the kink in the slope of the lines for small energies and the appearance of extended dark regions with vanishing conductance. Out of curiosity, we also show $G(\epsilon_F, B_{\mathrm{long}})$ for one single disorder configuration in Fig. 6.8 c), which shows a very complex structure. In order to explain the phenomena described above, further investigation is needed. Additional insight might be gained, for instance, by analyzing spin densities of scattering states. Unfortunately, the tight-binding simulations with discretized Dirac surface Hamiltonians do not grant reliable spin textures due to Fermion doubling (see Sec. 2.4.2). While a Wilson mass term directly alters the spin texture, *kwant* mixes physical and spurious solutions (which are degenerate) if no Wilson mass term is used. One possible remedy might be to manually separate physical from spurious solutions in the leads. Alternatively, it might be helpful to resort to full 3D simulations (including the bulk) or other methods which do not suffer from Fermion doubling.

7 Conclusions

In the course of this dissertation, we have investigated the magnetotransport properties of 3DTI nanowires when the Fermi energy lies in the bulk band gap, *i.e.* when transport is solely determined by topologically non-trivial surface states. We thereby focused on gated wires with constant cross section and on shaped nanowires. In the following, we first summarize our main findings, and then give an outlook toward possible future research directions.

In a joint experimental and theoretical effort, we investigated the surface states of strained HgTe nanowires [28]. The top gate used in these experiments induces a highly non-uniform electron density around the perimeter. We found that, despite the non-uniform electron density, the main transport features are captured by one *single* effective electron density. This observation relies on the fact that the surface electrons behave like ultrarelativistic particles and therefore Klein tunnel. By conducting a quantitative analysis of the conductance as a function of gate voltage, we were able to show that the surface states on the HgTe nanowires are not spin degenerate – which is a signature of their topological nature.

In the second part of this dissertation, we studied nanowires whose cross section varies along their length. We found that if rotational symmetry is preserved, magnetotransport properties are determined by a 1D Dirac equation with an effective mass potential accounting for the angular motion. This mass potential constitutes a powerful tool to predict the qualitative transport behavior of such nanowires without having to resort to numerical simulations.

Moreover, we found that a realistic (smoothed) truncated 3DTI nanocone attached to cylindrical leads is an extremely versatile system. Indeed, it allows access to wildly different transport regimes, simply by tuning and/or reorienting a homogeneous magnetic field:

i) At zero or weak magnetic field, it behaves like an ordinary quantum point contact.

ii) For stronger, coaxial magnetic fields, such that the magnetic length associated with the magnetic field component perpendicular to the surface is smaller than the length of the cone, *i.e.* $l_B \ll L$, transport is governed by resonant transmission through Dirac LLs.

iii) Once the magnetic field strength exceeds a certain threshold, determined by the QH states in the smoothed regions close to the leads and their disorder broadening, effective magnetic barriers emerge between the leads and the inner cone region. The latter thus becomes an electronic island, whose transport properties can be governed by Coulomb interactions between confined Dirac electrons. We dub such a system a "quantum magnetic bottle", in analogy to a classical magnetic bottle. Interestingly, confinement applies to all LLs except for the lowest one, whose energy is pinned to zero (due to its Dirac character) and thus immune to magnetic barrier formation.

iv) Rotating the magnetic field by $90°$ such that it is perpendicular to the nanowire axis, the cone enters an extrinsic higher-order TI phase with quantized transmission originating from topologically protected hinge states, provided the mangnetic length is smaller than the width of the wire.

Evidently, the truncated nanocone – which is realizable via state-of-the-art experimental capabilities – is an ideal platform to study a large variety of Dirac electron-related physics on the surface of 3DTIs.

In the last part of this dissertation, we devised a numerical tight-binding model with a varying grid constant along the hopping direction, which allows to model nanowires with broken rotational symmetry. As an example, we considered a nanowire constriction with rectangular cross section and constant height. The magnetoconductance features of this particular geometry together with all other results collected so far, allow us to make the following general statement: In perpendicular magnetic fields, when the local magnetic length is small compared to the width of the wire such that the side surfaces are separated by QH-gapped top and bottom surfaces, transport is invariably determined by chiral side surface states and the geometry is irrelevant for the qualitative conductance behavior. The reason for this is that the chiral side surface states do not feel the geometric shape of the nanowire due to their quasi 1D character. In coaxial and weak magnetic fields, however, states wrap around the wire circumference and the precise geometry has a crucial influence on the transport properties.

In the following, we discuss a number of open questions and thereby give an outlook toward possible future research directions:

i) Our description of Coulomb blockade in smoothed 3DTI nanocones in Sec. 6.2 is based on the constant interaction model (see Sec. 2.1), which we used to compute the addition spectra. On top of the addition spectra, it might also be of interest to compute the conductance, which can be done for given tunneling rates, single-particle energies, charging energy, and temperature [134]. Note, however, that computing the tunneling rates, which are determined by the effective mass potential barriers, is quite involved. The shape and height of the barrier depends on the magnetic field and on the precise geometry of the smoothed cone, and is different for each orbital angular momentum quantum number. For details, we refer to Ref. [133]. While the constant interaction model – assuming constant charging energy and single-particle energies which are not affected by the interactions – is reasonable as a first approximation, it would certainly be interesting to go beyond these assumptions. A review of more sophisticated models to treat transport through quantum dots in high magnetic fields can be found in Ref. [42]. Finally, note that there are other nanowire geometries of interest which act as a quantum magnetic bottle, as for instance a dumbbell geometry [133].

ii) As noted in Sec. 6.3, the research on the nanowire geometry which breaks rotational symmetry, sketched in Fig. 6.4, is not finished yet. Large parts of the rich transport signatures displayed in Fig. 6.8 are not understood. A possible remedy might be to resort to full 3D simulations using effective bulk Hamiltonians (see, for instance, Refs. [135, 136]), which allow to extract reliable spin textures of the scattering states. To this end, it might be helpful to utilize a non-uniform grid, similar to the one introduced in Sec. 6.3.1, in order to save computational resources by using fewer grid points deeper in the bulk.

iii) As mentioned briefly in a footnote in the beginning of Ch. 6, a cone has a point-like concentration of curvature at its apex referred to as conical singularity, which can have a profound influence on the electronic structure near the singular point. Such effects were studied for Schrödinger electrons in the QH regime in Ref. [127]. It was shown that the local excess angular momentum of a conical singularity is given by $L_{\mathrm{cone}}(\gamma) = \hbar c(1/\gamma - \gamma)/24$, where $\gamma = \sin(\beta/2)$ with the opening angle β, and the central charge $c = 1 - 12(1/2 - j\nu)/\nu$ with filling fraction ν and spin j. Interestingly, the excess angular momentum for a cusp, the singular point of a pseudosphere, is universally given by $L_{\mathrm{cusp}} = \hbar c/24$ [137]. In this regard two questions might arise: Is it possible to measure L_{cusp} in a real, *i.e.* finite system when appropriate limits are taken? Are there significant differences between Schrödinger and Dirac electrons when exposed to singular points of curvature?

iv) Recently, much attention has been devoted to *topological superconductivity*, which arises when topologically non-trivial materials and superconductivity are combined [9]. As an example, consider a TI nanowire proximity-coupled to an s-wave superconductor. It was shown that if the wire is subject to a coaxial magnetic field such that an odd number of modes is induced at the Fermi energy, Majorana zero modes are expected to emerge at its ends [138]. Such Majorana modes have intriguing properties which make them valuable for instance for potential applications in topological quantum computation. For details we refer to Refs. [139, 140]. The combination of shaped TI nanowires and superconductivity is, to the best of our knowledge, unexplored ground.

Let us conclude by emphasizing that TI nanowire junctions are a fascinating platform to explore the physics of 2D Dirac electrons in wildly different regimes. It is this versatility – together with the topological protection and the special properties of Dirac electrons – which makes such systems interesting not only for studying fundamental aspects of physics, but also for future technical applications.

Appendix

A.1 Schrödinger tight-binding Hamiltonian

The square of the momentum operator in the 1D Schrödinger Hamiltonian $\hat{H}_{\text{Schr}} = \hat{p}^2/(2m)$ can be discretized using the symmetric finite difference approach, which yields

$$\hat{p}^2\Psi(x)\Big|_{x=x_j} = -\frac{\hbar^2}{a^2}\left(\Psi_{j+1} - 2\Psi_j + \Psi_{j-1}\right), \tag{A.1}$$

where j is the lattice index and a the lattice constant. The Schrödinger equation $\hat{H}_{\text{Schr}}\Psi = \epsilon\Psi$ then defines a set of equations

$$\frac{\hbar^2}{2ma^2}\left(\Psi_{j+1} - 2\Psi_j + \Psi_{j-1}\right) = \epsilon\Psi_j, \tag{A.2}$$

which can be written in matrix form as

$$\begin{pmatrix} 2t & -t & 0 & 0 & & \dots & -t \\ -t & 2t & -t & 0 & & \dots & 0 \\ 0 & -t & 2t & -t & & \dots & 0 \\ \vdots & & & & & & \vdots \\ 0 & \dots & & & -t & 2t & -t \\ -t & \dots & & & 0 & t & 2t \end{pmatrix} \begin{pmatrix} \Psi_1 \\ \Psi_2 \\ \vdots \\ \vdots \\ \Psi_{N-1} \\ \Psi_N \end{pmatrix} = \epsilon \begin{pmatrix} \Psi_1 \\ \Psi_2 \\ \vdots \\ \vdots \\ \Psi_{N-1} \\ \Psi_N \end{pmatrix}, \tag{A.3}$$

where $t \equiv \hbar^2/(2ma^2)$. Here, we assume periodic boundary conditions which makes the matrix finite dimensional, and leads to the entries $-t$ in the upper right and the lower left corner. Using the Ansatz $\Psi_j = \frac{1}{\sqrt{L}}e^{ikja}$ in Eq. (A.2) yields the dispersion

$$\epsilon(k) = 2t\left[1 - \cos(ka)\right], \tag{A.4}$$

which corresponds exactly to the band of a simple atomic chain with one orbital and nearest-neighbor hopping.

A.2 Local spin rotation

In the following, we derive Eq. (3.2) with the unitary transformation $\hat{U}(\varphi) = \exp(-i\varphi\sigma_z/2)$ starting from Eq. (3.1). We will use the commutation relation for Pauli matrices $[\sigma_i, \sigma_j] = 2\mathrm{i}\sum_{k=1}^{3}\epsilon_{ijk}\sigma_k$, where ϵ_{ijk} is the Levi-Civita symbol, and we will use the Hadamard lemma

$$\mathrm{e}^X Y e^{-X} = \sum_{m=0}^{\infty} = \frac{1}{m!}[X,Y]_m, \tag{A.5}$$

with $[X,Y]_m = \left[X, [X,Y]_{m-1}\right]$, and $[X,Y]_0 = Y$. Using Eq. (A.5), we can write

$$\begin{aligned}
&\mathrm{e}^{\mathrm{i}\varphi\sigma_z/2}\sigma_z\mathrm{e}^{-\mathrm{i}\varphi\sigma_z/2} \\
&= \sigma_x + \mathrm{i}\frac{\varphi}{2}[\sigma_z,\sigma_x] + \frac{1}{2}\left(\frac{\mathrm{i}\varphi}{2}\right)^2[\sigma_z,[\sigma_z,\sigma_x]] + \frac{1}{3!}\left(\frac{\mathrm{i}\varphi}{2}\right)^3[\sigma_z,[\sigma_z,[\sigma_z,\sigma_x]]] + ... \\
&= \left(1 - \frac{\varphi^2}{2} + \frac{\varphi^4}{4!}...\right)\sigma_x - \left(\varphi - \frac{\varphi^3}{3!} + \frac{\varphi^5}{5!} - ...\right)\sigma_y \\
&= \cos(\varphi)\sigma_x - \sin(\varphi)\sigma_y,
\end{aligned} \tag{A.6}$$

where we used the Taylor expansions of sine and cosine. Analogously, one can show that

$$\mathrm{e}^{\mathrm{i}\varphi\sigma_z/2}\sigma_y\mathrm{e}^{-\mathrm{i}\varphi\sigma_z/2} = \cos(\varphi)\sigma_y + \sin(\varphi)\sigma_x. \tag{A.7}$$

The transformed Hamiltonian is given by

$$\tilde{H} = \hat{U}^{-1}H\hat{U} = \hbar v_F\left[k_z\sigma_z + \frac{1}{2}\left(\hat{U}^{-1}k_\varphi\sigma_\varphi\hat{U} + \hat{U}^{-1}\sigma_\varphi k_\varphi\hat{U}\right)\right] = \tag{A.8}$$

$$= \hbar v_F\left\{k_z\sigma_z - \frac{\mathrm{i}}{2R}\left[\hat{U}^{-1}(\partial_\varphi\sigma_\varphi)\hat{U} + 2\hat{U}^{-1}\sigma_\varphi(\partial_\varphi\hat{U}) + 2\hat{U}^{-1}\sigma_\varphi\hat{U}\partial_\varphi\right]\right\}, \tag{A.9}$$

where we omitted the φ dependence in $\hat{U}(\varphi)$ for convenience. As an example, let us compute the first term in the square brackets in Eq. (A.9), which yields

$$\begin{aligned}
\hat{U}^{-1}(\partial_\varphi\sigma_\varphi)\hat{U} &= \hat{U}^{-1}\left[-\cos(\varphi)\sigma_x - \sin(\varphi)\sigma_y\right]\hat{U} = \\
&= -\cos(\varphi)\hat{U}^{-1}\sigma_x\hat{U} - \sin(\varphi)\hat{U}^{-1}\sigma_y\hat{U} \\
&= -\sigma_x,
\end{aligned} \tag{A.10}$$

where we used Eq. (A.6) and (A.7). Using $\sigma_i\sigma_j = \delta_{ij}\sigma_0 + \mathrm{i}\sum_{k=1}^{3}\epsilon_{ijk}\sigma_k$, it can be shown that the second term in the square bracket in Eq. (A.9) yields σ_x, while the third term yields $2\sigma_y\partial_\varphi$. Putting everything together, we obtain the transformed Hamiltonian $\tilde{H} = \hbar v_F\left(\hat{k}_z\sigma_z + \hat{k}_\varphi\sigma_y\right)$.

A.3 Quantum capacitance

In this appendix, we introduce the concept of a quantum capacitance [141], and argue why it can be neglected in the analysis of the experiments published in Ref. [28] and discussed in Ch. 4. To this end, we extend the equation for the electrostatic energy of a gated system, given by Eq. (2.3), by a contribution from quantum single-particle energies ϵ_j. This yields

$$E(N, V_g) = \frac{(Ne)^2}{2C} - V_g Ne + \epsilon(N) \tag{A.11}$$

with $\epsilon(N) \equiv \sum_{j=1}^{N} \epsilon_j$, where the index j runs over the N lowest eigenenergies. For a given gate voltage V_g, the number of electrons within the gated region can be found by minimizing the above equation for integer N. The condition for the corresponding minimum is given by

$$\frac{1}{2}\left[E(N+1, V_g) - E(N-1, V_g)\right] = \frac{e^2}{C}N - eV_g + \epsilon_F \stackrel{!}{=} 0, \tag{A.12}$$

which yields $N = C(V_g - \epsilon_F/e)/e$. Here, we used that $[\epsilon(N+1) - \epsilon(N-1)]/2 = \epsilon_F$ at $T = 0$ by definition. Switching to the convention from Ch. 4, where capacitance is defined as charge *density* per voltage (with units $\mathrm{F/m^2}$), we obtain

$$n = \frac{C}{e}\left(V_g - \frac{\epsilon_F}{e}\right). \tag{A.13}$$

The above equation is the central object we use to compute the effect of the quantum capacitance on the electron density for a given gate voltage V_g. A similar equation can be found in Ref. [142], which deals with quantum capacitance corrections in multigated graphene sheets. Note that for simplicity, we assume that there is no initial doping, *i.e.* $n = 0$ for $V_g = 0$.[1]

Let us now elaborate on the meaning of Eq. (A.13). In Ch. 4, we use $n = CV_g/e$ to compute the electron density as a function of gate voltage. The additional term $-C\epsilon_F/e^2$ in Eq. (A.13) accounts for the fact that part of the gate voltage is used to pay for a shift of the Fermi energy, *i.e.* for increasing single particle energies. It is illuminating to express n in terms of ϵ_F in A.13, which can be done using Eq. (4.5) for 2D systems, and Eq. (4.4) for quasi-1D systems. For the 2D case (with $g_s = 1$), we have $n(\epsilon_F) = \epsilon_F^2/(4\pi\hbar^2 v_F^2)$ and thus

$$\frac{1}{4\pi\hbar^2 v_F^2}\epsilon_F^2 + \frac{C}{e^2}\epsilon_F - \frac{C}{e}V_g = 0, \tag{A.14}$$

[1]Initial doping n_0 with the corresponding energy ϵ_0 (such that $\epsilon_F = \epsilon_0$ for $V_g = 0$) is added by using $n = C\left(V_g - (\epsilon_F - \epsilon_0)/e\right)/e + n_0$ instead of Eq. (A.13).

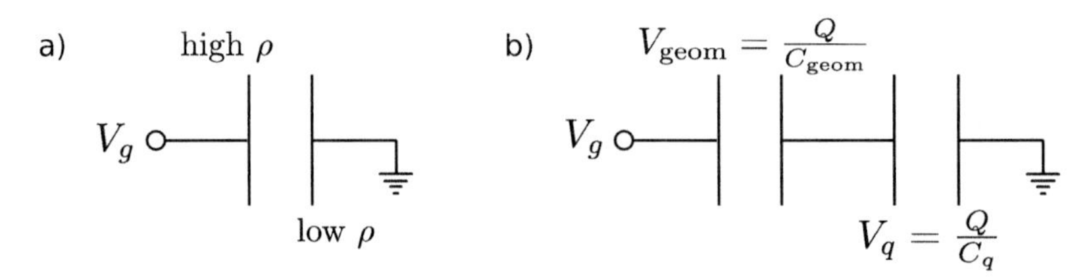

Figure A.1: a) Capacitor with a small density of states ρ on one of the plates. b) Series of two conventional capacitors representing the capacitor from a). Part of the gate voltage drops due to quantum corrections, which is incorporated in the quantum capacitance C_q (see main text for details).

which can be expressed in terms of the density of states $\rho(\epsilon_F) \equiv \mathrm{d}n/\mathrm{d}\epsilon|_{\epsilon=\epsilon_F}$ as

$$\left(2\rho(\epsilon_F) + \frac{C}{e^2}\right)\epsilon_F - \frac{C}{e}V_g = 0. \tag{A.15}$$

From the above equation, it can be concluded that the term C/e^2, originating from the quantum corrections, can be neglected if the density of states is large, *i.e.* if $2\rho(\epsilon_F) \gg C/e^2$. We will resort to this statement later on. Note that Eq. (A.14) can be directly used to determine the Fermi energy ϵ_F for a given gate voltage V_g, which can in turn be used to compute the electron density $n(\epsilon_F)$.

Let us briefly sketch a different approach to quantum capacitance in order to clarify its name. Here, we only aim at giving intuitive arguments; for a rigorous treatment, see for instance Refs. [141, 143]. We have seen in the last paragraph that a purely classical description of gating is insufficient as soon as the density of states is small enough such that the quantum correction term C/e^2 becomes important. For this reason, consider a plate capacitor where one of the plates has a small density of states (in the sense described in the previous sentence), as sketched in Fig. A.1 a). As explained before, part of the applied gate voltage drops due to an increase of the Fermi energy. In the following, we denote this part V_q due to its *quantum* nature. The remaining part V_{geom} pays for Coulomb repulsion, which is associated with the classical, *geometrical* capacitance. From a conceptual point of view, it is thus useful to split the applied gate voltage V_g such that $V_g = V_{\text{geom}} + V_q$. Hence, the equivalent circus to a plate capacitor where one plate has a small density of states are two capacitors in series, which is displayed in Fig. A.1 b). The voltage drop $V_{\text{geom}} = Q/C_{\text{geom}}$ is determined by the geometrical capacitance C_{geom}, which is obtained within the framework of classical electrostatics. The quantum capacitance C_q accounts for the voltage drop $V_q = Q/C_q$, and can be computed using

$$\mathrm{d}V_q = \frac{1}{e}\mathrm{d}\epsilon_F = \frac{1}{e}\mathrm{d}n\,\frac{\mathrm{d}\epsilon_F}{\mathrm{d}n} = \frac{1}{e}\frac{\mathrm{d}n}{\rho(\epsilon_F)}, \tag{A.16}$$

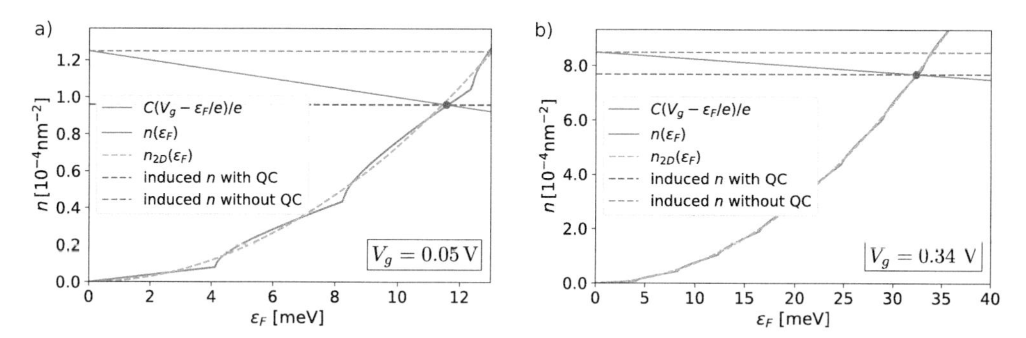

Figure A.2: Effect of quantum corrections on the electron density. See main text for details. a) For small gate voltages as V_g =0.05 V, the low density of states leads to large quantum corrections. b) A larger gate voltage $V_g = 0.34$ V renders quantum corrections less important. Note that the decisive parameter is actually the Fermi energy and not necessarily the gate voltage (important when initial doping is considered).

which yields

$$C_q = \frac{\mathrm{d}Q}{\mathrm{d}V_q} = e^2 \rho(\epsilon_F). \tag{A.17}$$

Returning to Eq. (A.13), we compute the effect of the quantum capacitance when gating an infinitely long cylindrical nanowire. We choose a realistic value for the geometrical capacitance $C_{\text{geom}} = 4 \times 10^{-4}\,\text{F/m}^2$ which matches one of the wires from Ref. [28]. This value is obtained using the numerical simulations discussed in Sec. 4.2. The resulting quantum corrections are presented in Fig. A.2. The blue curve represents the right hand side of Eq. (A.13), while the green (orange) curve corresponds to the 2D (quasi-1D) electron density as a function of Fermi energy [*i.e.* the left hand side of Eq. (A.13)]. The intersection of the curves (for the quasi-1D system marked with a red dot) yields the electron density (horizontal red line) with quantum corrections and the corresponding Fermi energy. The purple horizontal line represents the classical electron density (without quantum corrections). For $V_q = 0.05$ V [panel a)], the quantum corrections are clearly substantial: The electron density with quantum capacitance is around 23% smaller than the classically computed electron density. The reason is the small Fermi energy of 11.5 meV which results in a small density of states. For $V_g = 0.34$ V [panel b)], the Fermi energy is 32.4 V which results in a larger density of states, and thus comparable electron densities with an error below 10%.

As outlined in Sec. 4.4, one estimate yields 32 meV for the Fermi energy already at the valence band edge in 0.3% strained HgTe. This justifies the neglect of quantum corrections in Ch. 4 and Ref. [28].

A.4 Metric tensor on nanowire surfaces with rotational symmetry

The metric tensor g on the surface of the nanowire Σ with elements g_{ij} can be defined via the line element $\mathrm{d}l$, which is given by

$$\mathrm{d}l^2 = \sum_{ij} g_{ij} \mathrm{d}q_i \mathrm{d}q_j. \tag{A.18}$$

Here, i, j run over the coordinates r and φ. The tensor elements g_{ij} can be computed using

$$g_{ij} = \frac{\partial \boldsymbol{r}}{\partial q_i} \cdot \frac{\partial \boldsymbol{r}}{\partial q_j}, \tag{A.19}$$

where $\boldsymbol{a} \cdot \boldsymbol{b}$ denotes the scalar product between vector $\boldsymbol{a}$ and $\boldsymbol{b}$. Using Eq. 6.2, one readily obtains $g_{\varphi\varphi} = R^2(s)$ and $g_{s\varphi} = g_{\varphi s} = 0$. The tensor element g_{ss} is more involved and is derived in the following. Let us denote the derivatives of $R(s)$ and $z(s)$ as $R'(s) \equiv \mathrm{d}R/\mathrm{d}s(s)$ and $z'(s) \equiv \mathrm{d}z/\mathrm{d}s(s)$. Then, we can write

$$g_{ss} = \begin{pmatrix} R'(s)\cos\varphi \\ R'(s)\sin\varphi \\ z'(s) \end{pmatrix} \cdot \begin{pmatrix} R'(s)\cos\varphi \\ R'(s)\sin\varphi \\ z'(s) \end{pmatrix} = [R'(s)]^2 + [z'(s)]^2 \tag{A.20}$$

$$= \left(\frac{\mathrm{d}R}{\mathrm{d}s}\right) \left[1 + \left(\frac{\mathrm{d}z}{\mathrm{d}s}\frac{\mathrm{d}s}{\mathrm{d}R}\right)^2\right] = \left(\frac{\mathrm{d}R}{\mathrm{d}s}\right) \left[1 + \left(\frac{\mathrm{d}z}{\mathrm{d}R}\right)^2\right], \tag{A.21}$$

where we used that

$$s(R) = \int_{R_0}^{R} \mathrm{d}R' \sqrt{1 + \left(\frac{\mathrm{d}z}{\mathrm{d}R'}\right)^2} \tag{A.22}$$

can be locally defined on Σ. Using

$$\frac{\mathrm{d}s}{\mathrm{d}R} = \sqrt{1 + \left(\frac{\mathrm{d}z}{\mathrm{d}R}\right)} \tag{A.23}$$

we can write

$$g_{ss} = \left(\frac{\mathrm{d}R}{\mathrm{d}s}\right)^2 \left(\frac{\mathrm{d}s}{\mathrm{d}R}\right)^2 = 1. \tag{A.24}$$

With all the metric tensor elements, we can express the volume form as

$$\mathrm{d}V = \sqrt{|\det(g)|}\mathrm{d}\varphi \mathrm{d}s = R(s)\mathrm{d}\varphi \mathrm{d}s. \tag{A.25}$$

Using Eq. A.25, we can prove that the Dirac Hamiltonian on Σ, given by Eq. 6.4, is Hermitian, i.e. $\langle \Phi | H \Psi \rangle = \langle H \Phi | \Psi \rangle$ with the scalar product

$$\langle\Phi|\Psi\rangle \equiv \int_{-\infty}^{\infty} \mathrm{d}s \int_0^{2\pi} \mathrm{d}\varphi \; R(s)\Phi^*(s,\varphi)\Psi(s,\varphi). \tag{A.26}$$

We set $\hbar v_F = 1$ in the following and write

$$\langle\Phi|H\Psi\rangle = \int_{-\infty}^{\infty} \mathrm{d}s \int_0^{2\pi} \mathrm{d}\varphi \, R(s)\Phi^*(s,\varphi)(-\mathrm{i}) \left[\left(\partial_s + \frac{R'(s)}{2R(s)}\right)\sigma_z + \frac{1}{R(s)}\partial_\varphi\right]\Psi(s,\varphi). \tag{A.27}$$

We can pull the azimuthal derivative ∂_φ to the left using partial integration, *i.e.*

$$\int_{-\infty}^{\infty} \mathrm{d}s \int_0^{2\pi} \mathrm{d}\varphi \, R(s)\Phi^*(\varphi,s)(-\mathrm{i})\partial_\varphi\Psi(\varphi,s) \tag{A.28}$$

$$= \int_{-\infty}^{\infty} \mathrm{d}s \int_0^{2\pi} \mathrm{d}\varphi \, R(s)\left[-\mathrm{i}\partial_\varphi\Phi(\varphi,s)\right]^* \Psi(\varphi,s), \tag{A.29}$$

by assuming that $R(s)$ is defined for $-\infty < s < \infty$ and that the wave functions $\Phi(s,\varphi)$ and $\Psi(s,\varphi)$ vanish at infinity. The term with the derivative ∂_s reveals why the term originating from the spin connection is necessary to ensure Hermiticity. The derivative ∂_s acts on $R(s)$, which stems from the volume form, yielding

$$\int_{-\infty}^{\infty} \mathrm{d}s \int_0^{2\pi} \mathrm{d}\varphi \, R(s)\Phi^*(\varphi,s)(-\mathrm{i})\left(\partial_s + \frac{R'(s)}{2R(s)}\right)\Psi(s,\varphi) \tag{A.30}$$

$$= \int_{-\infty}^{\infty} \mathrm{d}s \int_0^{2\pi} \mathrm{d}\varphi \, \mathrm{i}\left\{\partial_s\left[R(s)\Phi^*(s,\varphi)\right] - \frac{1}{2}R'(s)\Phi^*(s)\right\}\Psi(s,\varphi) \tag{A.31}$$

$$= \int_{-\infty}^{\infty} \mathrm{d}s \int_0^{2\pi} \mathrm{d}\varphi \, \mathrm{i}\left[R(s)\partial_s\Phi^*(s,\varphi) - \frac{1}{2}R'(s)\Phi^*(s,\varphi)\right]\Psi(s,\varphi) \tag{A.32}$$

$$= \int_{-\infty}^{\infty} \mathrm{d}s \int_0^{2\pi} \mathrm{d}\varphi \, R(s)\left\{(-i)\left[\partial_s - \frac{R'(s)}{2R(s)}\right]\Phi(s,\varphi)\right\}^*\Psi(s,\varphi), \tag{A.33}$$

which proves Hermiticity.

Bibliography

[1] C. L. Kane and E. J. Mele, "Topological Order and the Quantum Spin Hall Effect", Phys. Rev. Lett. **95**, 146802 (2005).

[2] B. A. Bernevig, T. L. Hughes, and S.-C. Zhang, "Quantum Spin Hall Effect and Topological Phase Transition in HgTe Quantum Wells", Science **314**, 1757 (2006).

[3] M. Koenig, S. Wiedmann, C. Bruene, A. Roth, H. Buhmann, L. W. Molenkamp, X.-L. Qi, S.-C. Zhang, M. König, S. Wiedmann, C. Brüne, A. Roth, H. Buhmann, L. W. Molenkamp, X.-L. Qi, and S.-C. Zhang, "Quantum Spin Hall Insulator State in HgTe Quantum Wells", Science **318**, 766 (2007).

[4] K. v. Klitzing, G. Dorda, and M. Pepper, "New Method for High-Accuracy Determination of the Fine-Structure Constant Based on Quantized Hall Resistance", Phys. Rev. Lett. **45**, 494 (1980).

[5] M. Nakahara, *Geometry, topology and physics*, Bristol, UK: Hilger, Graduate student series in physics (1990).

[6] D. J. Thouless, M. Kohmoto, M. P. Nightingale, and M. den Nijs, "Quantized Hall Conductance in a Two-Dimensional Periodic Potential", Phys. Rev. Lett. **49**, 405 (1982).

[7] Enerdata, *Ict energy consumption*, `https://www.enerdata.net/publications/executive-briefing/expected-world-energy-consumption-increase-from-digitalization.html` (visited on 08/14/2019).

[8] J. E. Moore and L. Balents, "Topological invariants of time-reversal-invariant band structures", Phys. Rev. B **75**, 121306 (2007).

[9] M. Z. Hasan and C. L. Kane, "Colloquium: Topological insulators", Rev. Mod. Phys. **82**, 3045 (2010).

[10] J. H. Bardarson and J. E. Moore, "Quantum interference and Aharonov-Bohm oscillations in topological insulators", Rep. Prog. Phys. **76**, 056501 (2013).

[11] O. Klein, "Die Reflexion von Elektronen an einem Potentialsprung nach der relativistischen Dynamik von Dirac", Zeitschrift für Physik **53**, 157 (1929).

[12] D. Hsieh, D. Qian, L. Wray, Y. Xia, Y. S. Hor, R. J. Cava, and M. Z. Hasan, "A topological Dirac insulator in a quantum spin Hall phase", Nature **452**, 970 (2008).

[13] D. Hsieh, Y. Xia, L. Wray, D. Qian, A. Pal, J. H. Dil, J. Osterwalder, F. Meier, G. Bihlmayer, C. L. Kane, Y. S. Hor, R. J. Cava, and M. Z. Hasan, "Observation of Unconventional Quantum Spin Textures in Topological Insulators", Science **323**, 919 (2009).

[14] Y. L. Chen, J. G. Analytis, J.-H. Chu, Z. K. Liu, S.-K. Mo, X. L. Qi, H. J. Zhang, D. H. Lu, X. Dai, Z. Fang, S. C. Zhang, I. R. Fisher, Z. Hussain, and Z.-X. Shen, "Experimental Realization of a Three-Dimensional Topological Insulator, Bi_2Te_3", Science **325**, 178 (2009).

[15] J. G. Checkelsky, Y. S. Hor, M.-H. Liu, D.-X. Qu, R. J. Cava, and N. P. Ong, "Quantum Interference in Macroscopic Crystals of Nonmetallic Bi_2Se_3", Phys. Rev. Lett. **103**, 246601 (2009).

[16] N. P. Butch, K. Kirshenbaum, P. Syers, A. B. Sushkov, G. S. Jenkins, H. D. Drew, and J. Paglione, "Strong surface scattering in ultrahigh-mobility Bi_2Se_3 topological insulator crystals", Phys. Rev. B **81**, 241301 (2010).

[17] H. Peng, K. Lai, D. Kong, S. Meister, Y. Chen, X. L. Qi, S. C. Zhang, Z. X. Shen, and Y. Cui, "Aharonov-Bohm interference in topological insulator nanoribbons", Nature Materials **9**, 225 (2010).

[18] J. H. Bardarson, P. W. Brouwer, and J. E. Moore, "Aharonov-Bohm oscillations in disordered topological insulator nanowires", Phys. Rev. Lett. **105**, 156803 (2010).

[19] B. Hamdou, J. Gooth, A. Dorn, E. Pippel, and K. Nielsch, "Surface state dominated transport in topological insulator Bi_2Te_3 nanowires", Appl. Phys. Lett. **103**, 193107 (2013).

[20] S. S. Hong, Y. Zhang, J. J. Cha, X. L. Qi, and Y. Cui, "One-dimensional helical transport in topological insulator nanowire interferometers", Nano Lett. **14**, 2815 (2014).

[21] L. A. Jauregui, M. T. Pettes, L. P. Rokhinson, L. Shi, and Y. P. Chen, "Magnetic field-induced helical mode and topological transitions in a topological insulator nanoribbon", Nat. Nanotechnol. **11**, 345 (2016).

[22] S. Cho, B. Dellabetta, R. Zhong, J. Schneeloch, T. Liu, G. Gu, M. J. Gilbert, and N. Mason, "Aharonov-Bohm oscillations in a quasi-ballistic three-dimensional topological insulator nanowire", Nat. Commun. **6** (2015).

[23] J. Dufouleur, L. Veyrat, A. Teichgräber, S. Neuhaus, C. Nowka, S. Hampel, J. Cayssol, J. Schumann, B. Eichler, O. G. Schmidt, B. Büchner, and R. Giraud, "Quasiballistic Transport of Dirac Fermions in a Bi_2Se_3 Nanowire", Phys. Rev. Lett. **110**, 186806 (2013).

[24] F. Xiu, L. He, Y. Wang, L. Cheng, L. T. Chang, M. Lang, G. Huang, X. Kou, Y. Zhou, X. Jiang, Z. Chen, J. Zou, A. Shailos, and K. L. Wang, "Manipulating surface states in topological insulator nanoribbons", Nat. Nanotechnol. **6**, 216 (2011).

[25] Y. Zhang, Y. Ran, and A. Vishwanath, "Topological insulators in three dimensions from spontaneous symmetry breaking", Phys. Rev. B **79**, 245331 (2009).

[26] Y. Zhang and A. Vishwanath, "Anomalous Aharonov-Bohm conductance oscillations from topological insulator surface states", Phys. Rev. Lett. **105**, 2 (2010).

[27] K. Moors, P. Schüffelgen, D. Rosenbach, T. Schmitt, T. Schäpers, and T. L. Schmidt, "Magnetotransport signatures of three-dimensional topological insulator nanostructures", Phys. Rev. B **97**, 245429 (2018).

[28] J. Ziegler, R. Kozlovsky, C. Gorini, M. H. Liu, S. Weishäupl, H. Maier, R. Fischer, D. A. Kozlov, Z. D. Kvon, N. Mikhailov, S. A. Dvoretsky, K. Richter, and D. Weiss, "Probing spin helical surface states in topological HgTe nanowires", Phys. Rev. B **97**, 035157 (2018).

[29] M. Sitte, A. Rosch, E. Altman, and L. Fritz, "Topological insulators in magnetic fields: quantum Hall effect and edge channels with a nonquantized θ term.", Phys. Rev. Lett. **108**, 126807 (2012).

[30] J. Langbehn, Y. Peng, L. Trifunovic, F. von Oppen, and P. W. Brouwer, "Reflection-Symmetric Second-Order Topological Insulators and Superconductors", Phys. Rev. Lett. **119**, 246401 (2017).

[31] M. Geier, L. Trifunovic, M. Hoskam, and P. W. Brouwer, "Second-order topological insulators and superconductors with an order-two crystalline symmetry", Phys. Rev. B **97**, 205135 (2018).

[32] F. Schindler, A. M. Cook, M. G. Vergniory, Z. Wang, S. S. P. Parkin, B. A. Bernevig, and T. Neupert, "Higher-order topological insulators", Sci. Adv. **4**, 6 (2018).

[33] R. Landauer, "Spatial Variation of Currents and Fields Due to Localized Scatterers in Metallic Conduction", IBM Journal of Research and Development **1**, 223 (1957).

[34] R. Landauer, "Electrical resistance of disordered one-dimensional lattices", The Philosophical Magazine: A Journal of Theoretical Experimental and Applied Physics **21**, 863 (1970).

[35] R. Landauer, "Residual resistivity dipoles", Zeitschrift für Physik B Condensed Matter **21**, 247 (1975).

[36] M. Büttiker, "Scattering theory of thermal and excess noise in open conductors", Phys. Rev. Lett. **65**, 2901 (1990).

[37] M. Büttiker, "Scattering theory of current and intensity noise correlations in conductors and wave guides", Phys. Rev. B **46**, 12485 (1992).

[38] M. Buttiker, "Capacitance, admittance, and rectification properties of small conductors", J. of Phys. Condens. Matter **5**, 9361 (1993).

[39] S. Datta, *Electronic transport in mesoscopic systems*, Cambridge Studies in Semiconductor Physics and Microelectronic Engineering (Cambridge University Press, 1995).

[40] D. V. Averin and K. K. Likharev, "Coulomb blockade of single-electron tunneling, and coherent oscillations in small tunnel junctions", J. Low Temp. Phys. **62**, 345 (1986).

[41] H. van Houten and C. W. J. Beenakker, "Comment on "Conductance oscillations periodic in the density of a one-dimensional electron gas"", Phys. Rev. Lett. **63**, 1893 (1989).

[42] U. Meirav and E. B. Foxman, "Single-electron phenomena in semiconductors", Semicond. Sci. Technol. **11**, 255 (1996).

[43] L. P. Kouwenhoven, D. G. Austing, and S. Tarucha, "Few-electron quantum dots", Rep. Prog. Phys. **64**, 701 (2001).

[44] D. Averin and A. Odintsov, "Macroscopic quantum tunneling of the electric charge in small tunnel junctions", Phys. Lett. A **140**, 251 (1989).

[45] L. Glazman and K. Matveev, "Residual quantum conductivity under Coulomb-blockade conditions", Sov. Phys.–JETP Lett. **51**, 484.

[46] Glattli D. C., Pasquier C., Meirav U., Williams F. I. B., Jin Y., and Etienne B., "Co-tunneling of the charge through a 2-D electron island", Zeitschrift für Physik B Condensed Matter **85**, 375 (1991).

[47] P. Fulde, J. Keller, and G. Zwicknagl, "Theory of Heavy Fermion Systems", Solid State Phys. **41**, 1 (1988).

[48] J. M. Kosterlitz and D. J. Thouless, "Ordering, metastability and phase transitions in two-dimensional systems", J. Phys. C **6**, 1181 (1973).

[49] N. D. Mermin and H. Wagner, "Absence of Ferromagnetism or Antiferromagnetism in One- or Two-Dimensional Isotropic Heisenberg Models", Phys. Rev. Lett. **17**, 1133 (1966).

[50] F. D. M. Haldane, "Ground State Properties of Antiferromagnetic Chains with Unrestricted Spin: Integer Spin Chains as Realisations of the O(3) Non-Linear Sigma Model", (1981), arXiv:`1612.00076 [cond-mat.other]`.

[51] F. D. M. Haldane, "Continuum dynamics of the 1-D Heisenberg antiferromagnet: Identification with the O(3) nonlinear sigma model", Phys. Lett. A **93**, 464 (1983).

[52] F. D. M. Haldane, "Nonlinear Field Theory of Large-Spin Heisenberg Antiferromagnets: Semiclassically Quantized Solitons of the One-Dimensional Easy-Axis Néel State", Phys. Rev. Lett. **50**, 1153 (1983).

[53] B. Simon, "Holonomy, the Quantum Adiabatic Theorem, and Berry's Phase", Phys. Rev. Lett. **51**, 2167 (1983).

[54] M. V. Berry, "Quantal phase factors accompanying adiabatic changes", Proceedings of the Royal Society A, Mathematical and Physical Sciences **392**, 45 (1984).

[55] F. Bloch, "Über die Quantenmechanik der Elektronen in Kristallgittern", Zeitschrift für Physik **52**, 555 (1929).

[56] A. Altland and M. R. Zirnbauer, "Nonstandard symmetry classes in mesoscopic normal-superconducting hybrid structures", Phys. Rev. B **55**, 1142 (1997).

[57] A. P. Schnyder, S. Ryu, A. Furusaki, and A. W. W. Ludwig, "Classification of topological insulators and superconductors in three spatial dimensions", Phys. Rev. B **78**, 195125 (2008).

[58] F. J. Dyson, "The Threefold Way. Algebraic Structure of Symmetry Groups and Ensembles in Quantum Mechanics", J. Math. Phys. **3**, 1199 (1962).

[59] K. Shiozaki and M. Sato, "Topology of crystalline insulators and superconductors", Phys. Rev. B **90**, 165114 (2014).

[60] C.-K. Chiu, J. C. Y. Teo, A. P. Schnyder, and S. Ryu, "Classification of topological quantum matter with symmetries", Rev. Mod. Phys. **88**, 035005 (2016).

[61] K. S. Novoselov, A. K. Geim, S. V. Morozov, D. Jiang, Y. Zhang, S. V. Dubonos, I. V. Grigorieva, and A. A. Firsov, "Electric field effect in atomically thin carbon films", Science **306**, 666 (2004).

[62] C. L. Kane and E. J. Mele, "Z_2 Topological Order and the Quantum Spin Hall Effect", Phys. Rev. Lett. **95**, 146802 (2005).

[63] H. Min, J. E. Hill, N. A. Sinitsyn, B. R. Sahu, L. Kleinman, and A. H. MacDonald, "Intrinsic and Rashba spin-orbit interactions in graphene sheets", Phys. Rev. B **74**, 165310 (2006).

[64] Y. Yao, F. Ye, X.-L. Qi, S.-C. Zhang, and Z. Fang, "Spin-orbit gap of graphene: First-principles calculations", Phys. Rev. B **75**, 041401 (2007).

[65] L. Fu, C. L. Kane, and E. J. Mele, "Topological Insulators in Three Dimensions", Phys. Rev. Lett. **98**, 106803 (2007).

[66] R. Roy, "Topological phases and the quantum spin hall effect in three dimensions", Phys. Rev. B **79**, 195322 (2009).

[67] G. Tkachov, *Topological Insulators: The Physics of Spin Helicity in Quantum Transport* (CRC Press, 2015).

[68] L. Fu, "Hexagonal Warping Effects in the Surface States of the Topological Insulator Bi_2Te_3", Phys. Rev. Lett. **103**, 266801 (2009).

[69] Z. Alpichshev, J. G. Analytis, J.-H. Chu, I. R. Fisher, Y. L. Chen, Z. X. Shen, A. Fang, and A. Kapitulnik, "STM Imaging of Electronic Waves on the Surface of Bi_2Te_3: Topologically Protected Surface States and Hexagonal Warping Effects", Phys. Rev. Lett. **104**, 016401 (2010).

[70] K. Kuroda, M. Arita, K. Miyamoto, M. Ye, J. Jiang, A. Kimura, E. E. Krasovskii, E. V. Chulkov, H. Iwasawa, T. Okuda, K. Shimada, Y. Ueda, H. Namatame, and M. Taniguchi, "Hexagonally Deformed Fermi Surface of the 3D Topological Insulator Bi_2Se_3", Phys. Rev. Lett. **105**, 076802 (2010).

[71] W.-Y. Shan, H.-Z. Lu, and S.-Q. Shen, "Effective continuous model for surface states and thin films of three-dimensional topological insulators", New J. Phys. **12**, 043048 (2010).

[72] D. Culcer, E. H. Hwang, T. D. Stanescu, and S. Das Sarma, "Two-dimensional surface charge transport in topological insulators", Phys. Rev. B **82**, 155457 (2010).

[73] O. V. Yazyev, J. E. Moore, and S. G. Louie, "Spin Polarization and Transport of Surface States in the Topological Insulators Bi_2Se_3 and Bi_2Te_3 from First Principles", Phys. Rev. Lett. **105**, 266806 (2010).

[74] Z.-H. Pan, E. Vescovo, A. V. Fedorov, D. Gardner, Y. S. Lee, S. Chu, G. D. Gu, and T. Valla, "Electronic Structure of the Topological Insulator Bi_2Se_3 Using Angle-Resolved Photoemission Spectroscopy: Evidence for a Nearly Full Surface Spin Polarization", Phys. Rev. Lett. **106**, 257004 (2011).

[75] Y. Xia, D. Qian, D. Hsieh, L. Wray, A. Pal, H. Lin, A. Bansil, D. Grauer, Y. S. Hor, R. J. Cava, and M. Z. Hasan, "Observation of a large-gap topological-insulator class with a single dirac cone on the surface", Nat. Phys. **5**, 398 (2009).

[76] C. Brüne, C. X. Liu, E. G. Novik, E. M. Hankiewicz, H. Buhmann, Y. L. Chen, X. L. Qi, Z. X. Shen, S. C. Zhang, and L. W. Molenkamp, "Quantum Hall Effect from the Topological Surface States of Strained Bulk HgTe", Phys. Rev. Lett. **106**, 126803 (2011).

[77] L. Fu and C. L. Kane, "Topological insulators with inversion symmetry", Phys. Rev. B **76**, 045302 (2007).

[78] N. N. Berchenko and M. V. Pashkovskii, "Mercury telluride – a zero-gap semiconductor", Soviet Physics Uspekhi **19**, 462 (1976).

[79] S. C. Wu, B. Yan, and C. Felser, "Ab initio study of topological surface states of strained HgTe", EPL **107** (2014).

[80] J. Ziegler, "Quantum transport in HgTe topological insulator nanostructures", Dissertation (University of Regensburg, 2018).

[81] D. A. Kozlov, Z. D. Kvon, E. B. Olshanetsky, N. N. Mikhailov, S. A. Dvoretsky, and D. Weiss, "Transport Properties of a 3D Topological Insulator based on a Strained High-Mobility HgTe Film", Phys. Rev. Lett. **112**, 196801 (2014).

[82] R.-L. Chu, W.-Y. Shan, J. Lu, and S.-Q. Shen, "Surface and edge states in topological semimetals", Phys. Rev. B **83**, 075110 (2011).

[83] O. Crauste, Y. Ohtsubo, P. Ballet, P. A. L. Delplace, D. Carpentier, C. Bouvier, T. Meunier, A. Taleb-Ibrahimi, and L. Lévy, "Topological surface

states of strained Mercury-Telluride probed by ARPES", (2013), arXiv:1307.2008.

[84] C. W. Groth, M. Wimmer, A. R. Akhmerov, and X. Waintal, "Kwant: A software package for quantum transport", New J. Phys. **16**, 1 (2014).

[85] T. P. Santos, L. R. F. Lima, and C. H. Lewenkopf, "A numerical method to efficiently calculate the transport properties of large systems: an algorithm optimized for sparse linear solvers", (2018), arXiv:1812.07709.

[86] L. Susskind, "Lattice fermions", Phys. Rev. D **16**, 3031 (1977).

[87] R. Stacey, "Eliminating lattice fermion doubling", Phys. Rev. D **26**, 468 (1982).

[88] H. B. Nielsen and M. Ninomiya, "A no-go theorem for regularizing chiral fermions", Phys. Lett. B **105**, 219 (1981).

[89] H. B. Nielsen and M. Ninomiya, "Absence of neutrinos on a lattice: (I). Proof by homotopy theory", Nucl. Phys. B **185**, 20 (1981).

[90] H. B. Nielsen and M. Ninomiya, "Absence of neutrinos on a lattice: (II). Intuitive topological proof", Nucl. Phys. B **193**, 173 (1981).

[91] B. Messias De Resende, F. C. De Lima, R. H. Miwa, E. Vernek, and G. J. Ferreira, "Confinement and fermion doubling problem in Dirac-like Hamiltonians", Phys. Rev. B **96**, 161113 (2017).

[92] K. M. Habib, R. N. Sajjad, and A. W. Ghosh, "Modified Dirac Hamiltonian for efficient quantum mechanical simulations of micron sized devices", Appl. Phys. Lett. **108** (2016).

[93] P. Mello, P. Pereyra, and N. Kumar, "Macroscopic approach to multichannel disordered conductors", Annals of Physics **181**, 290 (1988).

[94] C. W. J. Beenakker, "Random-matrix theory of quantum transport", Rev. Mod. Phys. **69**, 731 (1997).

[95] J. H. Bardarson, J. Tworzydło, P. W. Brouwer, and C. W. J. Beenakker, "One-Parameter Scaling at the Dirac Point in Graphene", Phys. Rev. Lett. **99**, 106801 (2007).

[96] F. de Juan, J. H. Bardarson, and R. Ilan, "Conditions for fully gapped topological superconductivity in topological insulator nanowires", SciPost Physics **6**, 060 (2019).

[97] J. H. Bardarson and R. Ilan, "Transport in Topological Insulator Nanowires", 93 (2018), arXiv:1906.05192v1.

[98] A. Sinha and R. Roychoudhury, "Dirac Equation in (1 + 1)-Dimensional Curved Space-Time", Int. J. Theor. Phys. **33** (1994).

[99] K.-I. Imura, Y. Takane, and A. Tanaka, "Spin berry phase in anisotropic topological insulators", Phys. Rev. B **84**, 195406 (2011).

[100] S. Essert, "Mesoscopic Transport in Topological Insulator Nanostructures", Dissertation (University of Regensburg, 2015).

[101] J. Zierenberg, N. Fricke, M. Marenz, F. P. Spitzner, V. Blavatska, and W. Janke, "Percolation thresholds and fractal dimensions for square and cubic lattices with long-range correlated defects", Phys. Rev. E **96**, 062125 (2017).

[102] E. Akkermans and G. Montambaux, *Mesoscopic Physics of Electrons and Photons* (Cambridge University Press, 2007).

[103] S. Adam, P. W. Brouwer, and S. Das Sarma, "Crossover from quantum to Boltzmann transport in graphene", Phys. Rev. B **79**, 201404 (2009).

[104] T. Ando and H. Suzuura, "Presence of Perfectly Conducting Channel in Metallic Carbon Nanotubes", J. Phys. Soc. Japan **71**, 2753 (2002).

[105] J. Luttinger, "The Effect of a Magnetic Field on Electrons in a Periodic Potential", Phys. Rev. **84**, 814 (1951).

[106] L. Trifunovic and P. W. Brouwer, "Higher-Order Bulk-Boundary Correspondence for Topological Crystalline Phases", Phys. Rev. X **9**, 011012 (2019).

[107] D. H. Lee, "Surface States of Topological Insulators: The Dirac Fermion in Curved Two-Dimensional Spaces", Phys. Rev. Lett. **103**, 6 (2009).

[108] Y.-Y. Zhang, X.-R. Wang, and X. C. Xie, "Three-dimensional topological insulator in a magnetic field: chiral side surface states and quantized Hall conductance", J. Phys. Condens. Matter **24**, 015004 (2012).

[109] E. Xypakis and J. H. Bardarson, "Conductance fluctuations and disorder induced $\nu = 0$ quantum Hall plateau in topological insulator nanowires", Phys. Rev. B **95**, 35415 (2017).

[110] L. Brey and H. A. Fertig, "Electronic states of wires and slabs of topological insulators: Quantum Hall effects and edge transport", Phys. Rev. B **89**, 085305 (2014).

[111] E. J. König, P. M. Ostrovsky, I. V. Protopopov, I. V. Gornyi, I. S. Burmistrov, and A. D. Mirlin, "Half-integer quantum Hall effect of disordered Dirac fermions at a topological insulator surface", Phys. Rev. B **90**, 165435 (2014).

[112] K.-M. Dantscher, D. A. Kozlov, P. Olbrich, C. Zoth, P. Faltermeier, M. Lindner, G. V. Budkin, S. A. Tarasenko, V. V. Bel'kov, Z. D. Kvon, N. N. Mikhailov, S. A. Dvoretsky, D. Weiss, B. Jenichen, and S. D. Ganichev, "Cyclotron-resonance-assisted photocurrents in surface states of a three-dimensional topological insulator based on a strained high-mobility HgTe film", Phys. Rev. B **92**, 165314 (2015).

[113] A. Logg, K.-A. Mardal, G. N. Wells, et al., *Automated solution of differential equations by the finite element method* (Springer, 2012).

[114] C. Geuzaine and J.-F. Remacle, "Gmsh: a three-dimensional finite element mesh generator with built-in pre- and post-processing facilities", Int J Numer Methods Eng **79**, 1309 (2009).

[115] M. H. Liu, *National Cheng Kung University, Taiwan.*

[116] D. K. Cheng, *Field and wave electromagnetics* (Reading, Mass: Addison Wesley, 1983).

[117] P. E. Allain and J. N. Fuchs, "Klein tunneling in graphene: Optics with massless electrons", Eur. Phys. J. B **83**, 301 (2011).

[118] A. G. Aronov and Y. V. Sharvin, "Magnetic flux effects in disordered conductors", Rev. Mod. Phys. **59**, 755 (1987).

[119] M. Kessel, J. Hajer, G. Karczewski, C. Schumacher, C. Brüne, H. Buhmann, and L. W. Molenkamp, "CdTe-HgTe core-shell nanowire growth controlled by RHEED", Phys. Rev. Mat. **1**, 23401 (2017).

[120] M. Fecko, *Differential Geometry and Lie Groups for Physicists* (Cambridge university press, 2006).

[121] E. Xypakis, J.-W. Rhim, J. H. Bardarson, and R. Ilan, "Perfect transmission and Aharanov-Bohm oscillations in topological insulator nanowires with nonuniform cross section", Phys. Rev. B **101**, 045401 (2020).

[122] A. Rycerz, “Magnetoconductance of the Corbino disk in graphene”, Phys. Rev. B **81** (2010).

[123] Ş. Kuru, J. Negro, and L. M. Nieto, “Exact analytic solutions for a Dirac electron moving in graphene under magnetic fields”, J. Condens. Matter Phys. **21**, 455305 (2009).

[124] R. Kozlovsky, A. Graf, D. Kochan, K. Richter, and C. Gorini, “Magnetoconductance, Quantum Hall Effect, and Coulomb Blockade in Topological Insulator Nanocones”, (2019), arXiv:1909.13124.

[125] N. M. R. Peres, F. Guinea, and A. H. Castro Neto, “Electronic properties of disordered two-dimensional carbon”, Phys. Rev. B **73**, 125411 (2006).

[126] B. Dóra, “Disorder effect on the density of states in Landau quantized graphene”, Low Temp. Phys. **34**, 801 (2008).

[127] T. Can, Y. H. Chiu, M. Laskin, and P. Wiegmann, “Emergent Conformal Symmetry and Geometric Transport Properties of Quantum Hall States on Singular Surfaces”, Phys. Rev. Lett. **117** (2016).

[128] A. De Martino, L. Dell’Anna, and R. Egger, “Magnetic confinement of massless dirac fermions in graphene”, Phys. Rev. Lett. **98**, 1 (2007).

[129] C.-G. Fälthammar, “Effects of time-dependent electric fields on geomagnetically trapped radiation”, Journal of Geophysical Research (1896-1977) **70**, 2503 (1965).

[130] G. D. Reeves, H. E. Spence, M. G. Henderson, S. K. Morley, R. H. W. Friedel, H. O. Funsten, D. N. Baker, S. G. Kanekal, J. B. Blake, J. F. Fennell, S. G. Claudepierre, R. M. Thorne, D. L. Turner, C. A. Kletzing, W. S. Kurth, B. A. Larsen, and J. T. Niehof, “Electron Acceleration in the Heart of the Van Allen Radiation Belts”, Science **341**, 991 (2013).

[131] K. Storm, G. Nylund, L. Samuelson, and A. P. Micolich, “Realizing Lateral Wrap-Gated Nanowire FETs: Controlling Gate Length with Chemistry Rather than Lithography”, Nano Lett. **12**, 1 (2012).

[132] M. Royo, M. De Luca, R. Rurali, and I. Zardo, “A review on III-V core-multishell nanowires: Growth, properties, and applications”, J. Phys. D Appl. Phys. **50**, 143001 (2017).

[133] A. Graf, “Geometry-induced magnetic confinement and Coulomb blockade in shaped topological insulator nanowires”, Master Thesis (Universität Regensburg, 2019).

[134] C. W. J. Beenakker, "Theory of Coulomb-blockade oscillations in the conductance of a quantum dot", Phys. Rev. B **44**, 1646 (1991).

[135] H. Zhang, C.-X. Liu, X.-L. Qi, X. Dai, Z. Fang, and S.-C. Zhang, "Topological insulators in Bi_2Se_3, Bi_2Te_3 and Sb_2Te_3 with a single Dirac cone on the surface", Nat. Phys. **5**, 438 (2009).

[136] C.-X. Liu, X.-L. Qi, H. Zhang, X. Dai, Z. Fang, and S.-C. Zhang, "Model Hamiltonian for topological insulators", Phys. Rev. B **82**, 045122 (2010).

[137] T. Can and P. Wiegmann, "Quantum Hall states and conformal field theory on a singular surface", J. Phys. A-Math. Theor. **50**, 494003 (2017).

[138] A. Cook and M. Franz, "Majorana fermions in a topological-insulator nanowire proximity-coupled to an s-wave superconductor", Phys. Rev. B **84**, 1 (2011).

[139] C. Nayak, S. H. Simon, A. Stern, M. Freedman, and S. Das Sarma, "Non-Abelian anyons and topological quantum computation", Rev. Mod. Phys. **80**, 1083 (2008).

[140] J. Manousakis, A. Altland, D. Bagrets, R. Egger, and Y. Ando, "Majorana qubits in a topological insulator nanoribbon architecture", Phys. Rev. B **95**, 165424 (2017).

[141] S. Luryi, "Quantum capacitance devices", Appl. Phys. Lett. **52**, 501 (1988).

[142] M.-H. Liu, "Theory of carrier density in multigated doped graphene sheets with quantum correction", Phys. Rev. B **87**, 125427 (2013).

[143] T. Fang, A. Konar, H. Xing, and D. Jena, "Carrier statistics and quantum capacitance of graphene sheets and ribbons", Appl. Phys. Lett. **91**, 092109 (2007).

Acronyms

3DTI 3D topological insulator

ARPES angle-resolved photoemission spectroscopy

FFM Fourier filtering method

LL Landau level

LLL lowest Landau level

QD quantum dot

QH quantum Hall

QHE quantum Hall effect

QSHE quantum spin Hall effect

SOC spin-orbit coupling

TI topological insulator

TRS time-reversal symmetry

TSS topological surface states

Index

Acknowledgments

I would like to express my deep gratitude to everyone who has supported me in the course of this dissertation.

First and foremost, I would like to thank Klaus Richter for inviting me into his wonderful research group and supporting me in every possible way. Without his guidance and his creative ideas, this work would not have been possible. He always encouraged me to actively engage in the scientific life by giving me the opportunity to participate in numerous conferences and to spend valuable time in research facilities abroad.

Here, I would especially like to thank Ming-Hao Liu for inviting me to the National Cheng Kung University in Tainan. I felt extremely welcome in his research group, and it was incredibly stimulating to learn about all the customs and traditions of Taiwanese people. It was a truly exciting experience that I will never forget.

Also, I would like to thank Siddharth Parameswaran for giving me the opportunity to conduct research in his group at the University of Oxford for almost half a year. I learned so much during the numerous discussions with Sid, Jedediah H. Pixley and Aaron Friedman.

I am especially grateful to Cosimo Gorini, who supported me whenever I needed help. We spend endless hours discussing problems in physics and forging plans how to tackle them. Without his ideas this work would not have been possible, and without his jokes it would have been way less entertaining.

Next, I would like to thank my office mates Josef Rammensee and Benjamin Geiger for the great time and the valuable discussions. Josef took me in as a new system administrator, and he was my mentor from then on. He always put the problems of others above his own, and I am very grateful for all of his help. Also, Benni was always happy to discuss physics whenever I had a question. He helped me numerous times, and he was ready to sacrifice one or the other afternoon in order to solve one of my problems.

Special thanks go to Denis Kochan, my expert in differential geometry, and expert in telling jokes. The coffee round with him was truly entertaining. Moreover, I would like to express my gratitude to Sven Essert for his guidance and support in the beginning of my PhD. I would also like to thank Ansgar Graf for all the valuable discussions and for the exceeding motivation with which he has worked on our joint projects.

I am grateful to Johannes Ziegler and Dieter Weiss for the fruitful collaboration, and for the insight into the world and problems of an experimentalist I gained during our project.

For valuable discussions about physics I would especially like to thank Jacob Fuchs, Michael Barth, Phillipp Reck and Andreas Bereczuk. Moreover, I want to express my gratitude to Jacob, Michael, Vanessa Junk, and Tobias Frank, for helping me to cope with the daily system administrator work.

For proofreading parts of this dissertation, I am thankful to Cosimo, Jacob, Michael, Johannes, Ansgar, and Magdalena Marganska-Lyzniak.

I acknowledge funding from the Elite Network of Bavaria within the graduate school "Topological Insulators".

Finally, I am deeply thankful to my wife Marlene and to my family for always being there for me.